GALAPAGOS
AT THE CROSSROADS

GALAPAGOS
AT THE CROSSROADS

Pirates, Biologists, Tourists,
and Creationists Battle for
Darwin's Cradle of Evolution

CAROL ANN BASSETT

■ NATIONAL GEOGRAPHIC

WASHINGTON, D.C.

Published by the National Geographic Society

1145 17th Street N.W., Washington, D.C. 20036

ISBN: 978-1-4262-0402-9

The National Geographic Society is one of the world's largest nonprofit scientific and educational organizations. Founded in 1888 to "increase and diffuse geographic knowledge," the Society works to inspire people to care about the planet. It reaches more than 325 million people worldwide each month through its official journal, *National Geographic,* and other magazines; National Geographic Channel; television documentaries; music; radio; films; books; DVDs; maps; exhibitions; school publishing programs; interactive media; and merchandise. National Geographic has funded more than 9,000 scientific research, conservation and exploration projects and supports an education program combating geographic illiteracy. For more information, visit nationalgeographic.com.

For more information, please call 1-800-NGS LINE (647-5463) or write to the following address:

National Geographic Society
1145 17th Street N.W.
Washington, D.C. 20036-4688 U.S.A.

Visit us online at www.nationalgeographic.com

For information about special discounts for bulk purchases, please contact
National Geographic Books Special Sales: ngspecsales@ngs.org

For rights or permissions inquiries, please contact National Geographic Books Subsidiary Rights:
ngbookrights@ngs.org

Printed in the United States of America.

09/RRDC/1

Interior design: Cameron Zotter

To the children of the Galápagos.
May you teach your parents well.

Contents

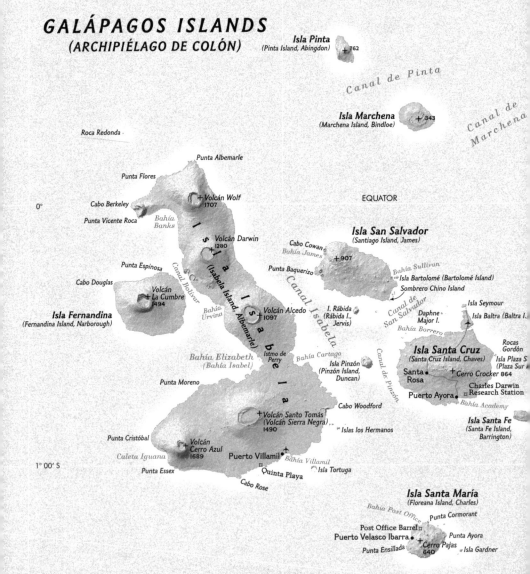

Isla Darwin (Darwin Island, Culpepper)

Isla Wolf (Wolf Island, Wenman)

P A C I F I C   O C E A N

1° 00' N

GALÁPAGOS ISLANDS
(ARCHIPIÉLAGO DE COLÓN)

Isla Pinta
(Pinta Island, Abingdon)
+ 762

Canal de Pinta

Isla Marchena
(Marchena Island, Bindloe)
+ 343

Canal de Marchena

Roca Redonda

Punta Albemarle

Punta Flores

0°

EQUATOR

Cabo Berkeley

+ Volcán Wolf
1707

Punta Vicente Roca

Bahía Banks

Volcán Darwin
1280

Isla San Salvador
(Santiago Island, James)

Cabo Cowan

Bahía James

Punta Espinosa

Punta Baquerizo

+ 907

Canal Bolívar

Isla Bartolomé (Bartolomé Island)

Cabo Douglas

Volcán
La Cumbre
1494

Isabela Island, Albemarle

Sombrero Chino Island

Volcán Alcedo
1097

I. Rábida
(Rábida I.,
Jervis)

Canal de San Salvador

Isla Fernandina
(Fernandina Island, Narborough)

Bahía Urvina

Daphne
Major I.

Isla Seymour

Isla Baltra (Baltra I.)

Bahía Borrero

Canal Isabela

Rocas
Gordón

Isla Santa Cruz
(Santa Cruz Island, Chaves)

Isla Plaza S
(Plaza Sur

*Bahía Elizabeth
(Bahía Isabel)*

Istmo de
Perry

Bahía Cartago

Isla Pinzón
(Pinzón Island,
Duncan)

Canal de Pinzón

Santa
Rosa

+ Cerro Crocker 864

Punta Moreno

Puerto Ayora

Charles Darwin
Research Station

Bahía Academy

Isabela

Cabo Woodford

+ Volcán Santo Tomás
(Volcán Sierra Negra)
1490

+ Islas los Hermanos

Isla Santa Fe
(Santa Fe Island,
Barrington)

Punta Cristóbal

+ Volcán
Cerro Azul
1689

1° 00' S

Caleta Iguana

Puerto Villamil

Bahía Villamil

Punta Essex

Quinta Playa

Isla Tortuga

Cabo Rose

Isla Santa María
(Floreana Island, Charles)

Bahía Post Office

Punta Cormorant

Post Office Barrel

Punta Ayora

Puerto Velasco Ibarra

+ Cerro Pajas
640

Punta Ensillada

Isla Gardner

92° 00' W

91° 00' W

Galápagos
Islands

ECUADOR   EQUATOR

SOUTH
AMERICA

Pacific
Ocean

Atlantic
Ocean

Isla Genovesa
(Genovesa Island, Tower)

ahía Darwin
arwin Bay)

◎ Provincial capital

Peak elevations are in meters.

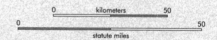

0   kilometers   50

0   50

statute miles

Archipiélago	Archipelago
Arrecife	Reef
Bahía	Bay
Cabo	Cape
Caleta	Cove, Inlet
Canal	Channel
Cerro	Peak
Isla–s	Island–s
Punta	Point
Roca–s	Rock–s
Volcán	Volcano

Bahía de Hobbs

Caleta Tortuga

Roca Pateadora
(León Dormido,   *Punta Dedo*
Kicker Rock)   Punta Pitt

Punta Bassa   **Isla San Cristóbal**
                                                    (San Cristóbal Island, Chatham)
Bahía   El Progreso
Baquerizo Moreno   Cerro San Joaquín **759**
(Wreck Bay)   ⊕ ┼ ■ —— *El Junco Lake*
                    Puerto Baquerizo Moreno

Bahía de Agua Dulce

hal de Santa Fe

P A C I F I C

O C E A N

Isla Española
(Española Island, Hood) *Isla Gardner*
Punta Suárez   ┼ **198**
                            Bahía Gardner
          Punta Cevallos

89° 00' W   88° 00' W

Introduction

In the time of the *garúa,* when cold Pacific currents flow north from Antarctica, then west along the Equator, an eerie mist shrouds the Galápagos Islands like a veil. In Spanish, garúa means "persistent light rain," and to the early explorers, pirates, and whalers who once passed through the archipelago, this ephemeral haze could be both seductive and deadly: The illusion of water called out like the enchanting Sirens of Greek mythology. Charles Darwin wrote in his field journal in September 1835: "The main evil under which these islands suffer is the scarcity of water." What Darwin and those before him did not understand was that the Galápagos had survived precisely because the lack of fresh water had rendered the islands uninhabitable to humans. But to the strange and lovely denizens dispersed to these islands by wind, water, and on rafts of organic debris, life had taken a different course. Land species that gained a foothold on these young volcanic islands adapted to whatever food was available, and they evolved in ways unimaginable. Marine species thrived in the plankton-rich waters, including penguins washed ashore by the Humboldt Current.

The Galápagos remained isolated until their accidental "discovery" in 1535 when Fray Tomás de Berlanga, the Bishop of Panama, and his crew, on a mission to what is now Peru, were caught in a dead calm and drifted westward on the ocean currents. Berlanga described the islands as "dross, worthless," and rejoiced when the trade winds returned to free his ship from this haunting inferno. In time, those who followed would learn to adapt to this volcanic landscape straddling the Equator, just as had the bizarre life-forms that evolved there throughout the millennia: giant land tortoises that can survive more than 150 years; marine iguanas able to hold their breath under water for up to an hour; cormorants that no longer need wings to fly; vampire finches that survive on the blood of masked boobies; daisies that have morphed into giant trees.

The Galápagos Islands and their unusual denizens, part of the nation of Ecuador since 1832, stand at a critical crossroads—on a collision course with 21st-century values driving tourism and immigration, and with invasive species that prey on the very life-forms that make the Galápagos special. In 2007 the United Nations Educational, Scientific, and Cultural Organization (UNESCO) declared the islands an endangered World Heritage site, and scientists, conservationists, and Galapagueños are working to protect the archipelago before it's too late.

Today, about 35,000 colonists call the Galápagos home, and more arrive every day, despite new laws intended to limit immigration. The population growth is 6 percent per year, compared with only 2 percent on the Ecuadorian mainland; demographers predict the population could well surpass 50,000 in the next few decades. Ninety-seven percent of the land mass of the islands is protected as a national park; the other 3 percent includes the towns and private land. Yet this 3 percent is where some of the worst problems originate. As the influx of settlers grows and the towns begin straddling the park's boundaries, social, economic, and environmental problems have multiplied. Invasive species brought to the islands over the centuries by colonists and tourists have wreaked havoc throughout the archipelago. Goats, burros, pigs, dogs, cats, and rats have decimated the wildlife and vegetation on several islands, including Isabela, where about 100,000 feral goats turned the verdant hills into a desert and nearly obliterated the giant land tortoises that lived on Volcán Alcedo. While eradication work has succeeded in eliminating those intruders on Isabela and other islands, it could take years before some islands return to their natural state, if ever.

I first visited the Galápagos in 1990 as a young journalist on assignment for a national magazine. Like most visitors, I felt that I was entering a primordial dreamscape where time and space had stood still. From the shore of Floreana Island, I watched a school of spotted eagle rays wing through the waves like butterflies. In the darkness of night, bottlenose dolphins frolicked beside our tour boat, their fins sparkling with bioluminescence—living light from microscopic organisms. I quickly realized that I had come to Las Encantadas (the Enchanted Islands), as the early Spanish explorers called the archipelago. In the skies above North Seymour Island male frigate birds floated on thermals, their throats puffed out like giant red balloons to attract mates. But what struck me most was the baby sea lion on Santiago Island that trailed me like a puppy and sniffed at my shoes, not knowing what to make of me. As I gazed one day out to sea I spotted a lone sea turtle plowing through the waves like a dark leviathan. Perhaps this is what Darwin saw when he wrote that in the Galápagos, "both in time and space, we seem to be brought near to that great fact—the mystery of mysteries—the first appearance of new beings on this earth."

But even back in 1990, signs of discord were already apparent everywhere I went. On Española Island one day, three large cruise ships arrived simultaneously. As many as 200 international visitors disembarked, all vying to glimpse a group of brine-snorting marine iguanas that had just emerged from the sea. As the tourists competed to photograph the reptiles, they also trampled the vegetation that supported a rare colony of waved albatross. This, to me, seemed more like Coney Island than an isolated rock out in the Pacific. Given the avalanche of visitors, our tour guides rushed us along so quickly that any opportunity for ecological interpretation of the wildlife and its habitat was altogether lost. As for the solitude one hopes to find in such a remote place, it was shockingly absent—except in rare moments, and then, observation led to insight.

If I was allowed to pause long enough, though, I could feel my own human connections to the earth, to the fire in stone that had created these islands in the first place. If I could squat on the beach and let sand sift through my fingers I might imagine the power of time. If I could tarry long enough to watch a blue-footed booby regurgitate fish to its young, I might grasp the fragile balance between life and death. These moments of solitude were rare.

The town centers were no less vexing. At the world-renowned Charles Darwin Research Station on Santa Cruz Island, many of us visitors were surprised to see the namesake creatures of the islands, giant Galápagos land tortoises, penned up in stone corrals. One of these captive animals, Lonesome George, the last of his species from Pinta Island, was off by himself, basking in the sun. Also confined at the research center were tortoises with names like César, Luís, Antonio, and Black Pirate. Naïvely, I had hoped to see these 600-pound reptiles in the wild, but to do so I would need a special Galápagos National Park permit to visit the misty highlands of Isabela Island, where the healthiest population lives on a remote volcano called Alcedo. Though I vowed to return, it would be nearly 20 years before that dream became a reality.

In 1990 an estimated 12,000 colonists lived on the islands, most of them in Puerto Ayora, whose streets were a mixture of rock and volcanic cinder quarried from the northern end of Santa Cruz Island. Fishermen in shorts, tattered shirts, and sandals worked in the equatorial sun, building wooden boats and weaving nets. Few amenities existed for visitors. Makeshift kiosks sold seafood and trinkets made of endangered black coral. The village consisted of a few modest hotels and a bar called El Booby whose T-shirts were popular among tourists. There was no bank, nor were there telephones. The electricity was turned off after midnight. None of the trendy art galleries, sushi restaurants, day spas, discotheques, or Internet cafés that now exist had ever been imagined. Today, when you ask a Galapagueño what time it is, most will pull out a cell phone to check. The truth is, the Galápagos, a region with no indigenous roots or mythology—a cluster of remote islands known to the world for less than five centuries—has gone from the Stone Age to the space age in a matter of decades.

The issue from early on has been a difference in values over what the Galápagos Islands represent. A scientific laboratory? A place to exploit the natural resources? A major destination for international tourism? Often, the three activities are at odds, and consensus among competing interests is hard to achieve. Meanwhile, many of the local resources have been indiscriminately plundered, especially the fisheries. The sea cucumber, for example, a bottom-feeder that filters waste and other nutrients that benefit many species, is prized as an aphrodisiac in Asia. In the 1990s the insatiable demand for this sausage-shaped invertebrate set off what some people called a modern-day "gold rush." Opportunists, like those in the mid-19th-century American West, poured in from the mainland. Many subsistence fishermen suddenly became wealthy; sea cucumbers can pull in as much money as cocaine on the international black market. Today, as a result of this short-lived bonanza, a subspecies of sea cucumber is seriously threatened. Those that survive are now found so deep in the ocean that many divers become sick or die trying to exploit what's left.

But life for these *pepineros*, as the divers are called, was good in the 1990s, and when the sea cucumber season began each year, hundreds of fishing boats dotted the seas throughout the archipelago. In only three months one year, park service officials estimated that as many as seven million sea cucumbers had been harvested, far more than the authorized annual limit of 550,000. Supply boats came from mainland Ecuador 600 miles to the east and set up illegal fishing camps stocked with groceries, street drugs, and beautiful young prostitutes. Many of these young women were placed in the local *chongos* (brothels), and sometimes they bartered for bags of the desiccated invertebrates, which they later traded for cash. On Isabela Island, a taxi driver named Danny told a colleague and me that the chongos are open only when the pepineros come to town, and each young woman is visited by 10 to 15 men per night. "It's good for the *chicas,*" he claimed. "Where else can they make that kind of money?"

Throughout the years officials have found numerous illegal fishing camps on national parkland. On the shores of Isabela and Fernandina Islands, unlicensed fishermen recently cut down and burned protected mangrove trees, home to the rare mangrove finch, to boil their sea cucumbers in brine and dry them on racks in the equatorial sun. They also slaughtered dozens of endangered land tortoises for food. Reacting to these violations one year, government officials shut down the fishing season a month early, spurring protests throughout the islands. Nevertheless illegal harvests continue, and a market still thrives for such oddities as Galápagos sea lion penises, which have joined the sea cucumber as a highly marketable form of "Viagra" in Asia.

Despite the political power of the fishermen, it is tourism that has run the islands since the 1960s and that brings in the most revenue. According to the Charles Darwin Foundation, more than 174,000 people visited the Galápagos in 2007, pouring in about $418 million a year, with more than $145.5 million in boats alone. Gross income from tourism has increased by an average of 14 percent each year, according to the foundation. The irony, and the biggest frustration among

Galapagueños, is that only a fraction of that money ever returns from the mainland to support the park or the municipalities, and no one knows exactly where the bulk of the income ends up. Most of the large touring companies contribute little directly to the local economy, and public services in some towns are almost nonexistent. One week while I was in Puerto Baquerizo Moreno, the provincial capital of Galápagos on San Cristóbal Island, a state of emergency was declared when the town ran out of water.

The irony is that while tourism helps the local economy, it also harms the resources. Tour ships routinely dump raw sewage and trash into the ocean, and diesel fuel leaks from ships' engines, even from those owned by self-described ecotourism companies. As the number of tourists grows, more ships and local tour boats are required. So is more fuel and lube oil. In 2001 a tanker, the *Jessica*, ran aground just off San Cristóbal Island after an inexperienced captain mistook a signal buoy for a lighthouse. The ship was carrying 160,000 gallons of diesel and 80,000 gallons of molasses-thick bunker oil to refuel a huge tour ship visiting the archipelago. Marine life and seabirds suffered. A few days later, Ecuador's president declared a state of emergency. It was a wake-up call directly tied to the boom in tourism.

Tour guides say that, today, more tourists come to the islands for recreation or photography than for a lesson in natural history. New plans continue to be hatched for diverse kinds of recreation as foreigners and Ecuadorians alike try to cash in on what they see as a piece of paradise: sportfishing for marlin at $10,000 a pop, sky-diving in groups of up to a hundred jumpers, and cruises on ships that can carry up to 500 passengers. Fishermen who have overfished the waters are now using their boats to haul tourists from island to island, and some are studying to become professional dive masters. Foreign sea kayak companies promise camping and close encounters with dolphins, sharks, penguins, and whales, and other outfitters promote big-league surfing on islands where the waves rival the best in Hawaii. In the summer of 2007, while I was teaching an environmental writing class in the Galápagos for the University of Oregon, authorities detained a ship carrying a mini-submarine. Its passengers had each paid $120,000 for an illegal sub-voyage through protected waters.

So it came as no surprise in April 2007 when Ecuadorian President Rafael Correa signed a decree that declared the islands "at risk" and "a national priority" for preservation. In June the same year UNESCO listed the Galápagos as an "endangered" World Heritage site. The irony is that almost 30 years earlier, in 1978, the Galápagos Islands became one of the first places on Earth ever designated for protection as a United Nations World Heritage site, also known as a World Patrimony—a gift to humanity.

Biological erosion on the islands has come from a history of political instability and corruption in Ecuador. Since 2002 there have been 14 directors of the Galápagos National Park, and the country has had seven presidents in the last ten years. Although the park has established legal protections, they are often overlooked or enforcement is thwarted by powerful interests, including unionized bureaucracies, which have their own priorities. "There are loopholes in the laws here in everything," said Diego Quiroga, a cultural anthropologist and co-director of the Galápagos Academic Institute for the Arts and Sciences (GAIAS) on San Cristóbal Island. "To tell you the truth," he said, "the system is very corrupt."

Change is inevitable on these islands, and with it comes a host of controversies that are rarely reported by the local or foreign media. The few journalists who do work in the Galápagos have little or no training in natural history or public affairs reporting. One of them, Carlos Macias, a broadcast journalist for many years, told my class at GAIAS that access to information is extremely difficult. "Politics and the economy are very difficult to cover because they're tied together," he said. "The media here are owned by powerful political parties."

In 1998 Ecuador passed the Special Law for Galápagos to give residents of the archipelago more autonomy within their province. As applied and interpreted, this law lessened restrictions on development, granted more fishing permits to locals, and in some years allowed larger quotas on daily catches of such sensitive species as sharks, lobsters, and sea cucumbers. It also granted more tourism permits to the locals despite a general lack of training. Although the law was well intentioned, it was weakly enforced, and in many ways it has backfired. In 2000, when a quota was placed on fishing for spiny lobsters, several hundred angry fishermen attacked Galápagos National Park facilities and the Charles Darwin Research Station. They harassed tourists, closed roads, destroyed park property and records, and took endangered land tortoises hostage. "You want war?" a fisherman from the town of Villamil who helped lead the siege recently told me in an interview. "We'll give you war. Revolution!" An uneasy truce was reached only after troops were called in from the mainland. This wasn't the first time local fishermen had revolted, and it will likely not be the last. Many Galápagos fishermen have become politicians, and they wield strong political clout throughout the islands and on the mainland.

While the Special Law for Galápagos provided more employment for the locals, it also relaxed the restrictions on tourism. An indirect effect of the law was lowering the qualifications that a Galapagueño needs to become a naturalist guide. The new law also eliminated guides who aren't permanent residents, including many of the most experienced. Top guides have told me they can earn $5,000 a month or more. In a nation where the annual per capita income is only about $4,500, this work is lucrative indeed. But here's the catch: Before the law was passed, park guides were required to be well educated in biology, geology, and the general ecology of the islands. They were also required to be fluent in at least two languages. Under the new law, the only requirements for a guide are a high school diploma, a few weeks of basic training, and a rudimentary understanding of English. There is also a new trend in this cradle of evolution: Some creationists—biblical literalists who believe Earth is only 6,000 years old—have become guides, as I discovered unexpectedly during a trip to Bartolomé Island in 2006 and again in 2008, when I spoke with several Jehovah's Witnesses, one of whom works as a naturalist guide for Galápagos National Park and does not believe in evolution.

With such disparate beliefs among local residents, I wondered how do Galapagueños themselves find a niche in a bioregion that is now endangered. One problem is that most residents are relative newcomers without a deep connection to the archipelago or its resources. The directors of the world-renowned Charles Darwin Foundation cite the "frontier mentality" among colonists and several factors that contribute to this attitude: First, most of the population has lived in the Galápagos for less than six years, having come from completely altered environments on the mainland—agricultural areas where the land has been cleared, or where dams have changed rivers, or where forests have been clear cut. Second, residents have virtually no access (without a guide) to the pristine areas of the national park, so they're unlikely to learn anything about their new home by direct experience. Third, the school system is so deficient that it recently cut funding for environmental education. Fourth, as the numbers of colonists and their progeny increase, the town centers have become too cramped for new housing; in addition, resources as basic as potable water and electricity, always inadequate to the demand, are now seriously taxed. Fifth, health care is almost nonexistent. The HIV rate hovers at about 15 percent, and tuberculosis is widespread on Santa Cruz Island. Alcohol abuse is prevalent in the towns, and marijuana and cocaine are readily available.

Then, too, there's a land ownership problem. In May 2008 an Ecuadorian attorney representing dozens of Galapagueños attempted a Marxist-style land grab on Santa Cruz Island, claiming farmland in the highlands that had been abandoned. This did not sit well with the provincial governor of the Galápagos, Eliécer Cruz, who declared: "This government will not permit invasions."

But change is in the wind. Literally. One of the most stunning projects I witnessed on San Cristóbal Island was a wind farm, its sleek white turbines spinning ghostlike in the misty highlands. Here, during the windy season, three windmills can produce up to 80 percent of electrical demand for the nearby provincial capital, Puerto Baquerizo Moreno. The wind farm has greatly reduced the use of fossil fuels and the production of greenhouse gases. A second wind facility is now planned for Baltra Island, which will reduce carbon dioxide emissions there by more than 5,000 tons per year. Recycling is now mandatory in the provincial capital, and residents who don't comply are subject to fines. Also, on Santa Cruz Island, engineers are planning to convert from diesel-spewing taxis and motorcycles to electric vehicles within the next decade. Organic farming is now being practiced and taught to about 50 farmers in the Santa Cruz highlands. And activist groups such as the Sea Shepherd Conservation Society are working hand in hand with the Galápagos National Park and the Galápagos Environmental Police to catch poachers in the protected marine reserve.

As for invasive species, about 100,000 feral goats were recently eradicated from Isabela Island by aerial sharpshooters, specially trained dogs, and "Judas" goats from New Zealand. Today, a habitat once ravaged by the goats has reverted to dense jungle, where tortoises are so fat they can barely move, as I finally saw on a trek up Volcán Alcedo in 2008. Throughout the islands, scientists are using GPS (global positioning system) to monitor sensitive marine species, such as Pacific green sea turtles, sharks, and lobsters. Geolocators, tiny light-capturing gadgets attached to the legs of albatross, make it possible to chart the migrations of these critically endangered birds as they circle the globe to feed. Recent DNA testing has revealed that about 25 percent of waved albatross aren't monogamous, as had been thought. Researchers are also using DNA to study the origins and health of everything from Darwin's finches to the giant land tortoises and the Galápagos penguins.

Ironically, many locals regard these scientists as the "enemy," foreigners who parachute in with fat funding and who care more about marine iguanas than about people. On a planet where the human population is hovering at nearly seven billion, establishing a balance with nature presents a dilemma worldwide: How do people and nature coexist? The problem is especially acute in a bioregion as fragile as the Galápagos. What happens when the human struggle for survival is stronger than conserving life forms that most Galapagueños, quite literally, have never seen nor heard of? These are some of the tough questions that have captivated me during my many visits to the islands.

In December 2007 I moved to the town of Puerto Ayora on Santa Cruz Island and rented a little bungalow just off Academy Bay. It's a private house, surrounded by a white stucco wall. Its garden is graced with an almond tree and a feathery acacia with blossoms the color of fire. An alarm clock isn't needed: Every morning at dawn, yellow warblers whistle down through the trees. The patio consists of black volcanic cinder mined from a cone on the other side of the island. From the terrace upstairs I can watch magnificent frigate birds circle the bay, pirating sardines from each other in mid-flight. Blue-footed boobies dive beak first into the turquoise waves, their bodies stretched out like daggers. During mating season, marine iguanas sometimes emerge from the bay and waddle like toy-size dinosaurs into the street, forcing taxi drivers on Avenida Charles Darwin to slam on their brakes. In the morning I can sit in my yard and watch lava lizards do push-ups, or welcome Darwin's finches, so unfazed by my presence that they alight on the table and peck at my toast.

My house has tile floors, a beamed ceiling, and oval windows whose wooden frames open out into the yard. Their odd design means they can't be screened. But it was scorching hot in December, with humidity to match, so they remained wide open. I must admit I have had a lifelong aversion to mosquitoes: Their stings welt into the size of a dime and the itching can last up to ten days. To my horror, during my first few weeks in town, mosquitoes invaded my house and seemed bent on devouring me. Then the first rains came, and with them an onslaught of other insects. Tiny ants marched single file up my walls, hauling the lacy green wing of a beetle. Cockroaches the size of B-52 bombers emerged from nowhere. When darkness fell, flying ants entered through my unscreened windows by the hundreds, attracted to the lights. They didn't sting, so I watched in fascination as whole colonies clacked around inside my rice-paper lanterns. When they began dive-bombing my computer and its lighted keyboard, I freaked. I ran to the kitchen and grabbed a can of odorless insecticide that the owner of the home had left beneath the sink. As I blasted away, winged ants fell to the floor in heaps. I plopped on the couch with my head in my hands, surveying the carnage. Then, looking up, I saw a small gecko emerge from a crack in the wall. I tried to shoo it away, but it was too late. The round-toed reptile had just nabbed a few ants with its lightning quick tongue before vanishing, probably to die of liver failure.

What had I done, and how was I supposed to behave as a member of *Homo sapiens* in the world-famous Galápagos Islands? I'd been here less than a month and was already at war with nature. The irony is that all these unwelcome "house guests" were *invasive* species brought here from elsewhere, and they now pose one of the greatest threats to the islands. But wasn't I also an intruder?

To protect my computer equipment, and my sanity, I installed two small air-conditioning units in my house. I justified this carbon footprint by convincing myself that my lifestyle in the Galápagos was more benign than my living pattern back home in Oregon. In Puerto Ayora I had no car; I walked or rode a bicycle. Nor did I have an oven, microwave, iron, dishwasher, fireplace, washer, dryer, or Jacuzzi. Even so, I had joined the ranks of those who had failed to adapt to this so-called garden of Eden.

With air-conditioning, the mosquitoes no longer entered my house. This was good: Dengue fever had arrived in the Galápagos a few years earlier, and medical experts say it's only a matter of time before West Nile virus and Avian flu arrive. There is also the threat of canine distemper, a disease that can jump species from dogs to sea lions. In 2001 canine distemper killed most of the dogs in the Galápagos but did not affect pinnipeds.

As I looked deeper into these issues, I asked myself on a daily basis: Are the Galápagos really more special than other places? Or are they one example among many microcosms that exist on this fragile planet we call home? I had to conclude that they are unique. Scientists have now said farewell to the Holocene and have rung in a new epoch. They've dubbed it the Anthropocene—a human-dominated age in which urban-industrial society has contributed to global warming, mass extinctions, the displacement of species and cultures, and the depletion of nonrenewable resources. The impacts, they say, are permanent; the course of evolution itself has been thrust into the great unknown.

The Galápagos Islands now stand at a critical crossroads: To heal and endure as one of the world's most intact natural museums, or to lose most of their biodiversity to human encroachment, just as the islands of Hawaii and Guam have. As longtime naturalist guide and dive master

Mathias Espinosa told me one day on Isabela Island, "This is our last chance to live in harmony with nature. If nature loses this battle then our species—*Homo sapiens*—is condemned to pack our backpacks and live on the moon."

CHAPTER 1

Genesis

*"Blue, green, grey, white, or black; smooth, ruffled,
or mountainous; that ocean is not silent."*

—H. P. Lovecraft

In the beginning, before the Galápagos Islands pushed up through the Pacific Ocean floor, they were formless potential. The ocean was quiet, a tourmaline world that mirrored the sky. For millions of years the future islands stewed in a subterranean purgatory near the ocean's floor. Then something shifted when a giant piece of the Earth's crust known as the Nazca plate passed over a hot spot in the planet's mantle. So much heat escaped from the Earth's core that liquid rock vented up through the waves, forming volcanoes one steaming inch at a time. These nascent volcanoes rose into the equatorial sky, eventually becoming mountains. Pressure mounted from within and ripped through their flanks, spewing ash hundreds of feet into the sky and launching volcanic bombs—lava that cooled into solid rock before ever hitting the ground. Rivers of fire snaked down canyons at temperatures up to 2000°F. As the magma cooled, some of it froze into mirror-like images of the rippling sea. On some islands the land morphed into hellish fields of razor-sharp lava, "as though God had showered stones," wrote Fray Tomás de Berlanga, the Bishop of Panama, whose accidental "discovery" of the islands helped put them on the map in 1535.

Exactly three centuries later in 1835, a young Charles Darwin described the Galápagos Islands like this: "Nothing could be less inviting than the first appearance … a broken field of black basaltic lava, thrown into the most rugged waves and crossed by great fissures." Robert FitzRoy, captain of the H.M.S. *Beagle,* upon which Darwin sailed, was a bit more concise: "A shore fit for pandemonium" was the entry he made in the ship's log. To the earliest visitors and those who have followed, this primordial world harbored some of the strangest creatures on Earth: reptiles that plunged from cliffs into the sea to feed on algae; land tortoises the size of a barrel; hawks so "tame" they could be swatted off a tree branch. The Galápagos Islands and their inhabitants would shift long-established paradigms of life on Earth and how it originated. They would also tweak the perceptions of Victorian society, its scientists, and especially its clerics, who held that God had created the world only 6,000 years earlier, and that *Homo sapiens* was placed on Earth in physical bodies exactly like those of modern humans.

The Galápagos are among the youngest and most geologically active archipelagos on Earth. The islands, which straddle the Equator about 600 miles west of Ecuador, were formed less than five million years ago. In geological time, that's a nanosecond. The archipelago contains 13 major islands and 6 smaller ones. More than 40 islets have names, but scores of smaller islets and rocks, you could call them, remain nameless. The youngest islands are located on the western edge of the archipelago. Isabela Island is by far the largest at nearly 1,800 square miles. Shaped like a giant seahorse, it has five active volcanoes, one of which erupted in 2005 and another in 2008. The youngest island is Fernandina, a single volcano about 700,000 years old.

The oldest islands lie along the eastern edge of the archipelago, where a much different landscape unfolds. San Cristóbal is the easternmost island—two volcanoes fused together that are no longer active.

The island contains the largest freshwater lake in the Galápagos, El Junco, which is believed to be the remnant of an extinct caldera. Santa Cruz Island, in the middle of the archipelago, supports one of the most diverse plant communities in the Galápagos, including a wet zone in the highlands where ferns, mosses, mushrooms, orchids, and bromeliads thrive in the garúa mist. Española Island, to the southeast of Santa Cruz, is the oldest, at about 3.4 million years, and one of the most stunning. Its tall cliffs support nesting grounds for the waved albatross, masked and blue-footed boobies, frigate birds, and mockingbirds so bold they'll drink from a cup. But Española, like the other eastern islands, is doomed.

The Galápagos Islands are moving east on the back of the Nazca plate at the alarming rate of about three inches a year. In time, as they ride this underwater train, they'll sink and pass beneath the South American continent. Not long ago, oceanographers discovered what are probably the first Galápagos islands ever formed: subterranean volcanoes, now extinct, that were part of the archipelago about 90 million years ago. But change here is a constant, and as the older islands die in the east, new islands are born over the ever present hot spot to the west: Evolution is carried in every single rock.

Oceanographers have long known that life exists at the bottom of even the deepest seas, and in 1977 humans visited an area northeast of the archipelago called the Galápagos Rift in a deep-sea submersible. What they found at a depth of 8,500 feet was a world so bizarre it might have come from another planet—life-forms thriving where no human had ever visited. At this depth, huge vents heat the water to temperatures of 650°F. Here, under intense pressure and in complete darkness, oxygen barely exists. Poisonous hydrogen sulfide dominates. Surprisingly, carbon dioxide is also present. When combined with intense heat, these gases first gave life to primitive forms of bacteria. Mats of this heat-loving bacteria lie attached to the ocean floor, surviving the most extreme conditions imaginable and providing food for hundreds of species, including clams, huge colonies of mussels, crabs, octopuses, and giant tubeworms that sway like daisies over the super-heated vents. Then, in 2002 the world's top oceanographers traveled to the area to investigate. They confirmed what biologists had discovered 25 years earlier: that life on Earth may have originated three to four billion years ago in hydrothermal waters on the ocean's floor when the seething planet was still an infant. It is possible that the Galápagos Rift and its life-giving vents could provide more information about the origin of species than Darwin ever imagined.

———————————

The Galápagos are oceanic islands, meaning they were never connected to a continent. That's part of what makes them special. Their isolation in time and space has created patterns of evolution that are radically different from those on the continents. Such far-out worlds often provide a rare window into the diversity of life and how it adapts under some of the most extreme conditions imaginable: gigantism in tortoises as large as wine barrels and in a daisy that mutated into a tree; mutation in a bird's wings that have made it flightless; sexual crossover in sea turtles whose nests can produce all females if the incubation chamber is too warm.

How these denizens of the archipelago arrived there, where they came from, and how they learned to adapt are what make the Galápagos one of the most intriguing natural laboratories on Earth. It's believed that over time, plants and animals accidentally reached the islands by water or air. Seeds were carried on the wind, on the wings or feet of birds, or in their bellies. Flamingos likely flew in from the Caribbean. Favorable currents could have carried sea lions south from as far away as California, and penguins swam north through the frigid waters of Antarctica aided by the powerful Humboldt Current. Land iguanas and other reptiles may have floated on rafts of organic debris spewed out as flotsam when rivers on the mainland crested their banks.

Over time, the species that arrived in the Galápagos learned to adapt to this harsh environment. Some changed subtly, while others evolved and adapted so radically on different islands that endemic species were born—unique entities found nowhere else on Earth. The archipelago is vast enough and the islands far enough apart that most species eventually found their own biological niches. In that way, they avoided competition and evolved into distinct subspecies in a scientific process called adaptive radiation.

The biodiversity of the Galápagos is very low compared with that of mainland Ecuador, but the islands are characterized by one of the highest levels of endemism in the world. About 560 species of plants are native to the islands, meaning they arrived by natural means. About one-third are endemic, including species of Galápagos cotton, tomato, pepper, guava, passionflower, and a member of the daisy family that has evolved into a giant tree. Two species of cactus are also endemic: the lava cactus and the candelabra cactus. The islands are home to about 58 bird species. About half are endemic. These include the Galápagos hawk, dove, flycatcher, waved albatross, red-billed tropicbird, yellow warbler, mockingbird, and 13 species of finch. Among the finches, some eat seeds or leaves, some use twigs or cactus spines to remove insect larvae from trees. One species on Wolf Island, known as the vampire finch, pecks at the bodies of masked or Nazca boobies and drinks their blood.

The waters surrounding the islands are home to about 400 species of fish. Fifty are endemic, including the Galápagos shark. Only six species of mammals live on the islands: two species of sea lion, two species of bat, and two species of rice rat. In this lost world, reptiles, not mammals, rule. Of the 27 kinds of reptiles, 17 are endemic. The most famous is the giant land tortoise, with 14 subspecies, 3 of which are now extinct. Some time long ago, a population of giant land iguanas evolved into seafaring lizards that plow through the waves.

Part of what makes the Galápagos Islands so unique is their location on the map—and their climate. The islands are bathed by a complex system of ocean currents. The Humboldt, or Peru, Current carries a huge volume of cold water up from the Antarctic. When it passes Peru it curves and joins the westward-flowing Equatorial Current, sending cool water into the archipelago and creating a dry, moderate climate. The abundance of life also depends on upwellings, which occur when ocean currents and trade winds combine and spread seawater out like a fan. To compensate for this spreading phenomenon, extra water must rise to the surface, or upwell. Deep ocean water is rich in nutrients created when marine life dies, sinks, and decomposes. These nutrients stimulate the growth of algae or phytoplankton, the base of the ocean's food chain. When the upwellings flourish, so does the marine life of the Galápagos.

The islands have two seasons: The dry or cool season from July to December is also known as the garúa season, during which a perpetual mist hugs the highlands. The wet or hot season extends from January to June. Both seasons are susceptible to climate change, especially the event called El Niño, when torrential rains flood the islands and heat the nutrient-rich waters. In 1982–83, an El Niño event nearly devastated local marine life. Penguins, fur seals, and marine iguanas died in record numbers. Seabirds could not raise their young. But land creatures prospered. Then in 1997–98, another El Niño hit the Galápagos. Sea lions and other species died by the score, and beaches became bone-heaps of carcasses rotting in the sun. At the same time, all that rain turned the normally dry lowlands into an emerald kingdom where birds and land-dwelling creatures thrived. The situation becomes even more complicated when weather patterns reverse. La Niña, the opposite of El Niño, occurs when water and air temperatures become cooler than normal. This phenomenon benefits marine species but puts terrestrial wildlife at risk—life caught in a precarious balance, when even a few degrees' difference can lead a species to extinction.

The unusual climate and ocean currents have played a major role in the evolution of Galápagos life-forms. Quite by accident, they also etched on the map a primeval world that remained isolated for millions of years.

CHAPTER 2

Discovery

"The real voyage of discovery consists not in seeking new landscapes but in having new eyes."

—Marcel Proust

Bishop Tomás de Berlanga of Panama had good intentions in 1535 when he and his men set sail for what is now Peru. His mission: to investigate conquistador atrocities among indigenous peoples after the fall of the Inca Empire. Berlanga planned to meet with Francisco Pizarro to promote peace and to help administer the silver and gold mines on land stolen from the Indians for King Carlos V of Spain. But Berlanga's ship, part of the royal Spanish Armada, was becalmed during the journey south. Its sails sagged beneath a hellish sun. Its water barrels were nearly empty. Six days later strong currents seized the ship and carried it west to a strange new world where ebony islands rose into mist. Berlanga, dressed in his black cassock, stood on the bow, the wind in his face, and prayed for deliverance. Then he lowered the dinghy and sent his sailors ashore to search for water and grass for their horses. "Once out," he later wrote to the king, "they found nothing but seals, and turtles, and such big tortoises that each could carry a man."

The men sailed on, to a larger, more promising island, dug a well, and found water "saltier than that of the sea." Dehydration set in and they began eating prickly pear cactus pads. "Although not very tasty," the bishop wrote, "we began to eat of them, and squeeze them to draw all the water from them, and drawn, it looked like slops of lye, and they drank it as if it were rose water." After saying Mass on Passion Sunday, Berlanga sent the men out in teams to search for water. Miraculously, in a rocky ravine, they found enough sweet water to fill eight hogsheads onboard. Finally, the winds picked up. The black-robed bishop thanked God for his mercy and set sail back to the mainland with barely enough water to survive.

Some people believe that, based on pottery sherds found in the Galápagos, Inca cultures might have visited the islands around 1485. Archaeological evidence has revealed that some of the sherds belonged to the Chorrera culture; others to members of the Valdivia or Manteño-Huancavilca peoples, who were well known for their seafaring prowess. But the consensus remains that the first humans to set foot on the Galápagos Islands were Bishop Tomás de Berlanga and his faithful sailors. The rest is history.

In 1546, fierce ocean currents swept a ship captained by another Spaniard, Diego de Rivadeneira, into the seas off the Galápagos as the bearded captain and his armor-clad men fled from rival conquistadores in Peru. They suffered from thirst, just as the bishop had, and they too survived to tell of their strange adventures. Rivadeneira dubbed the Galápagos the Islas Encantadas (Enchanted Islands) because they would appear then disappear into the mist like phantoms. The Spaniards also dubbed the islands Galápagos (*galápago:* archaic for "saddle"), to commemorate the saddleback tortoise whose carapace had adapted to allow the giant reptiles to extend their necks and feed on tall stands of prickly pear cactus. In the 1600s, the Galápagos became a refuge for pirates who preyed on towns on the mainland, and on Spanish and Portugese galleons loaded with gold.

William Dampier was no ordinary pirate. When he wasn't out plundering towns or treasure-laden ships, this long-haired opportunist from

England was writing in his journal or sketching, with remarkable accuracy, the strange life-forms he encountered on his journeys. Dampier was the first man to sail around the world three times. He first visited the Galápagos in 1679 and returned two more times. In 1684, he anchored at Buccaneer Cove on Santiago Island with the buccaneers Capt. John Cook and his navigator William Ambrose Cowley, who drew the first credible map of the islands. The pirates had arrived on a Danish ship they'd seized and renamed the *Bachelor's Delight*.

Today, Dampier is considered one of the most famous "literary pirates" of all time—a paradoxical gentleman who was more interested in science and the sensuality of nature than in the spoils of conquest. Yet few people have ever heard of Dampier, whose most vivid observations come from the Galápagos. What a world the pirate had found, a place where spiked lizards snorted brine and giant tortoises roamed the volcanic rocks. He described tortoise flesh as "so sweet that no pullet eats more pleasantly" and wrote that the fat of the reptiles could be rendered into oil to make dumplings. He coined and defined the word "subspecies" and noted that the green sea turtles of the Galápagos seemed a "bastard" variety of those he'd seen in the Caribbean. Dampier was onto something, but the world was not yet ready to embrace his hints at the possibility of evolution. The dominant belief, foisted on society in 1658 when Dampier was still quite young, was that the world was created in six days and that life as we know it began precisely at 9 a.m. on October 23, 4004 B.C.

Back in England, Dampier married when he was in his late 20s, but he left his bride, Judith, soon after the wedding and did not see her again for 12 years. Nor did he write about her in his journals, reserving his passion for nature. He preferred instead to observe the mating habits of the wildlife he encountered, especially that in the Galápagos: marine iguanas that expended so much energy diving for algae that nature had blessed them with two penises; the pouches of male frigate birds inflated like big red balloons to attract females; the tenacity of mating sea turtles while competing males sometimes piled up on the pair four at a time. His keen curiosity, his provocative prose, and his analytical mind resulted in three best-selling travel books, so eloquent that they transformed the literary landscape. Samuel Taylor Coleridge, whose epic poem *The Rime of the Ancient Mariner* rocked the English-speaking world, praised Dampier's "exquisite mind" and urged aspiring writers to imitate him. Daniel Defoe contemplated Dampier's voyages. Jonathan Swift was deeply influenced as well and even mentions him in *Gulliver's Travels*. The American novelist Herman Melville visited the Galápagos Islands as a young whaler and later wrote *Moby-Dick*, based in part on the real-life tragedy that befell the U.S.S. *Essex* during the War of 1812. Dampier's impressive body of work on trade winds, storms, and ocean currents stirred the imagination—and scientific insights—of Alexander von Humboldt, who called him "the finest of all travel writers." A young Charles Darwin brought his books along on the expedition of the H.M.S. *Beagle* more than a century after Dampier's death.

By the late 1790s whalers had replaced the pirates when a British captain named James Colnett visited the Galápagos looking for sperm whales. Colnett constructed a "post office" barrel on Floreana, one of the few islands with natural springs, where British and American sailors could leave letters for ships bound for home. The abundance of whales in the archipelago set off a frenzy, and dozens of ships sailed in every year from the United States and England, killing whales for their oil to light lamps, fur seals for their beautiful pelts, and so many tortoises that some subspecies soon became extinct. It's estimated that during the 19th century, as many as 200,000 tortoises were slaughtered. Some species of whales, too, were nearly extinguished.

During the War of 1812, when the United States went to war with Britain over trade restrictions, an American naval captain named David Porter commanded the gunship U.S.S. *Essex*. One of his missions was to halt British shipping in the eastern Pacific. In the spring of 1813, this astute officer sailed straight to the post office barrel on Floreana Island, raided its contents, and gathered much needed intelligence for his campaign. He and his men destroyed most of Britain's whaling fleet in the Galápagos. A year later, in 1814, the *Essex* succumbed to an attack by two British warships at Valparaíso, Chile, and Porter was forced to surrender. Porter, however, had made some important contributions. In addition to limiting the British whaling fleet, he charted the islands and documented some of their unusual wildlife, including the variety in the shells of the tortoises on the various islands. But, like the pirates and whalers before him, his presence had changed the islands forever: He released some of the first invasive species on the islands—goats, so they could graze—but the ungulates soon disappeared into the canyons and began to reproduce.

Later visitors released goats intentionally as a food source for visiting sailors. In time, hundreds of thousands of feral goats dominated Santiago, Isabela, and other islands, destroying the flora that the giant tortoises depended on for survival. Eventually, settlers brought dogs, cats, burros, pigs, and other animals, which also became feral and preyed on endemic life-forms. Today, although a feral goat eradication program on Isabela and other islands has succeeded, invasive species, including plants, insects, reptiles, and even birds, remain one of the greatest threats to the survival of Galápagos biodiversity.

In 1832 the fledgling Republic of Ecuador claimed the Galápagos Islands from Spain, which had shown little interest in this daunting outpost. The following year Ecuadorian officials decided to settle the islands. They called upon José Villamil, an Ecuadorian general who had lived in the Louisiana Territory before the United States purchased it from France. Villamil was sent to the Galápagos from the Ecuadorian mainland to found a colony on Floreana, where the life-giving springs made it possible to farm the rich volcanic soil. Villamil raised fruits, vegetables, cattle, pigs, and goats, and he traded with whalers passing through the archipelago.

When Charles Darwin arrived on Floreana in 1835, he counted between 200 and 300 residents. Many were convicts from the mainland, sent to labor in the sun. "They are nearly all people of colour, who have been banished for political crimes from the Republic of Ecuador," Darwin wrote in his journal. He also wrote:

> The inhabitants, although complaining of poverty, obtain without much trouble, the means of subsistence. In the woods there are many wild pigs and goats; but the staple article of animal food is supplied by the tortoises. Their numbers have of course been greatly reduced in this island, but the people yet count on two days' hunting giving them food for the rest of the week. It is said that formerly single vessels have taken away as many as seven hundred, and that the ship's company of a frigate some years since brought down in one day two hundred tortoises to the beach.

By 1846, the tortoises of Floreana were gone. About 2,000 free-ranging cattle had taken their place, along with the livestock the colonists had abandoned when the colony failed. Then in 1869, an officer named Manuel Cobos established a settlement on San Cristóbal Island with the labor of prisoners from the mainland. Like Villamil, he treated them like slaves in the new village he named Progreso (Progress). But Cobos was more Neanderthal than progressive. Deranged with power, he often staged public floggings and death by dismemberment. It is said that he raped women, invoking the ancient rite of *prima nocte,* in which the feudal lord takes the place of the groom and deflowers the bride on the first night of her marriage. But Cobos was a successful entrepreneur. On the fertile soil of San Cristóbal, his subjects raised cattle, fruit, and sugarcane. Cobos sold meat and leather from the cattle, fruit from the orchards, and fish harvested from the surrounding ocean. He also sold oil from the giant tortoises that were native to the island. But his tyranny intensified, and in 1904 one of his most loyal men hacked him to death with a machete. Quid pro quo.

In 1893, Antonio Gil established yet another penal colony on Isabela Island. He named it Puerto Villamil, after the general who was first sent to the Galápagos in 1832. Gil's prisoners mined coral and burned it to produce lime. They built a second village called Santo Tomás and raised cattle on the slopes of Volcán Sierra Negra. As late as the 1950s, three convict camps were still operating on Isabela Island. One, not far from Puerto Villamil, stands as testimony to the cruel colonization of the islands. Known as the Muro de las Lágrimas (Wall of Tears), it was one of the camps built by about 300 prisoners during the 1940s and 1950s. Their task: to construct their own concentration camp within a wall 328 feet high and 23 feet long made of sharp volcanic rocks stacked without mortar. Many of the convicts died of heat stroke or after falling down the steep slopes while heaving boulders or by gunshot under *la ley de fuga* (the law of flight) that allowed guards to shoot escaping prisoners without compunction. The penal colony was closed after a revolt in 1959, but its Wall of Tears lives on in infamy.

CHAPTER 3

The Eminently Curious Darwin

*"The human species is, in a word, an environmental abnormality.
Perhaps a law of evolution is that intelligence usually extinguishes itself."*

—E. O. Wilson

When Charles Darwin visited the Galápagos in 1835, he found the natural history of the islands "eminently curious." Here were species so odd they could be found nowhere else on Earth. Darwin concluded in his classic travel narrative *The Voyage of the Beagle:* "The archipelago is a little world within itself." He saw it as a fantastic wonderland where giant tortoises munched on cactus and sooty marine iguanas dove into the sea. He described an individual of the latter species as "a hideous-looking creature, of a dirty black colour, stupid and sluggish in its movements."

Darwin was 26 years old, a lanky man who stood nearly six feet tall. An amateur naturalist, he of course wanted to know more about these reptiles, so bending down he grabbed one by the tail and flung it far out into the surf. To his dismay, the iguana swam back and returned exactly to where he stood. Astonished by its apparent lack of fear, Darwin again tossed the iguana into the waves, and again it crawled back. "I several times caught this same lizard," he wrote, "and as often as I threw it in, it returned in the same manner." Darwin speculated that "this apparent stupidity" probably meant the reptile had no land predators (at the time) and had developed a "hereditary instinct" for survival.

Darwin, the man who would eventually become known as the father of evolution, was onto something that would rock the world as much as Copernicus had three centuries earlier when he proposed his theory that the Earth revolves around the sun and not the other way around. Darwin would shift the paradigm about how life evolves on Earth and turn Victorian society on its pointed head.

Charles Darwin was born into a wealthy family in Shrewsbury, Shropshire, England, on February 12, 1809, the same day Abraham Lincoln entered the world in the green woods of Kentucky. His grandfather, Erasmus Darwin, was a physician and a prominent intellectual who harbored early ideas about evolution. His father, Robert, was also a doctor and wanted the family legacy to live on through Charles. But a different destiny beckoned, one that Charles's overbearing father could not have imagined.

Ironically, as a child, Darwin was a lazy, unfocused boy, a slow learner who was homeschooled for a while by his sister, Caroline. When he was nine years old his father sent him to a boy's boarding school across the river from the family estate. Darwin found little joy studying ancient history, the classics, and especially Greek. But he loved reading Shakespeare and Lord Byron and books about faraway lands. He was an excellent runner and liked showing off his physical skills, perhaps to compensate for his lack of intellectual prowess.

It's likely that his interest in nature arose when he played in his father's gardens and while fishing along the River Severn behind his parents' house. But his greatest joy came during the summer hikes he took along the northern coast of Wales, where he collected rocks, seashells, and insects. Back home in Shrewsbury, Charles spent time helping his brother, Erasmus, construct experiments in his homemade chemistry lab, which drew ridicule from his classmates who nicknamed him "Gas Darwin." The family feared he would blow up the house.

When Charles was 16 his father yanked him out of the boys' school for being indolent and for receiving poor grades. Robert told his son that if he didn't straighten up, "You will be a disgrace to yourself and all your family." That summer Darwin helped his father in his medical lab, and in the fall he entered medical school at the University of Edinburgh in Scotland. He still loved chemistry, but the sight of blood made him queasy. Meanwhile, his interest in the natural world began to deepen. He learned taxidermy from a freed slave who had lived in Guyana, South America, and who lived nearby. Darwin longed to visit such exotic lands. He read books about natural history and kept a detailed field notebook. And he joined the Plinian Society, which focused more on the natural world than a realm created by divine intervention. Then he met a zoology professor named Robert Grant, who taught Darwin how to collect and dissect marine organisms.

Sometimes Grant and his protégé took long walks at the seashore, discussing the intricacies of biology. It was here that the esteemed professor introduced Darwin to the ideas of the French naturalist and zoologist Jean-Baptiste Lamarck, who could be called the real father of evolution. In the 1700s Lamarck published books on zoology and paleontology, in effect creating a new field of biology. He believed it was the environment that caused changes in species and that parents passed these new traits down to their offspring. Darwin didn't know what to make of these revolutionary ideas. Like most Europeans of his time (including Lamarck) he was a creationist, a biblical literalist who believed that God had created the world in seven days and had designed human beings just as they appear today.

Although Darwin's imagination had been stirred by many new ideas, he still fared poorly in medical school. After two years he returned in shame to his father's house, where this time the patriarch decided his listless son would become a country minister. So Darwin attended Cambridge University at Christ's Church, where he was more interested in collecting beetles than in listening to sermons. Rev. John Henslow, a professor and naturalist, saw potential in Darwin and took him under his wing. Henslow held Friday dinner parties for upperclassmen, and on those evenings he spoke eloquently on matters of science. Darwin was his favorite student, and the two walked all over Cambridge, discussing geology, botany, entomology, math, and chemistry. "The man who walks with Henslow," as Darwin became known, had finally found his niche and with it new zeal for the wonders of science. Soon he began studying with one of the most renowned geologists in all of England, Adam Sedgwick. His lessons in field geology would prove invaluable, especially when Darwin found himself on volcanic islands like the Galápagos.

In 1831 Henslow helped Darwin join the survey ship H.M.S. *Beagle* on a journey that would last almost five years. Darwin signed on as an amateur naturalist and gentleman companion to Capt. Robert FitzRoy, who loved to engage in intellectual discourse. FitzRoy was four years older than Darwin and was often quite temperamental. The captain believed that a person's character could be judged by the shape of his face. When he first met Darwin he was somewhat jarred, for he observed in Darwin's nose a lack of energy—laziness. Nevertheless, the young naturalist won the commission and joined the expedition. Although he was seasick during most of the journey, his shipmates loved him and called him "the Philosopher."

Darwin's relationship with FitzRoy was not as smooth, and in fact the two men often disagreed. For one thing, the captain was more dogmatic in his Victorian views of creation than his gentleman companion. When Darwin boarded the *Beagle* at age 22, he brought with him a small library, including Milton's *Paradise Lost* (he loved poetry), the works of Alexander von Humboldt on oceanography and climate, and the best-selling books by the 18th-century literary pirate and naturalist William Dampier, who had visited the Galápagos more than a century earlier. Darwin also included a volume of Charles Lyell's *Principles of Geology:* The book's thesis states that Earth was formed over a much longer time than the Old Testament preached. To FitzRoy this was sheer blasphemy, and Darwin avoided the subject at all costs.

When the *Beagle* reached the Galápagos in September 1835 Darwin was struck first by the weird geology of the archipelago. On San Cristóbal Island he camped one night in an area where he counted 60 reddish craters and lava blown into "great bubbles." Darwin rose early the next day and explored the island with notebook in hand:

> The day was glowing hot, and the scrambling over the rough surface and through the intricate thickets, was very fatiguing; but I was well repaid by the strange Cyclopean scene. As I was walking along I met two large tortoises, each of which must have weighed two hundred pounds: one was eating a piece of cactus, and as I approached, it stared at me and slowly walked away; the other gave a deep hiss, and drew in its head. These huge reptiles, surrounded by the black lava, the leafless shrubs, and large cacti, seemed to my fancy like some antediluvian animals.

A few days later the *Beagle* sailed to Floreana Island, where a small penal colony had been established about six years earlier. Vice-Governor Nicholas Lawson, an Englishman, greeted the ship and told Darwin to observe the differences in the shapes of tortoise carapaces across the islands. Lawson said that by observing that feature, he had learned to determine, with certainty, which island a tortoise had come from. A tortoise with a tall saddle-shaped carapace, for example, came from an island where the primary food source was tall prickly pear cactuses. Lower-necked species could more easily find nourishment along the ground. This important clue did not sink in, and Darwin's failure to heed Lawson's advice would later haunt him. He did, however, observe how tortoises mated, what they ate, and how they drank water. He even rode them like ponies. "I was always amused when overtaking one of these great monsters, as it was quietly pacing along, to see how suddenly, the instant I passed, it would draw in its head and legs … I frequently got on their backs, and then giving a few raps on the hinder part of their shells, they would rise up and walk away: but I found it very difficult to keep my balance."

Like the pirates and whalers before him, Darwin also *ate* the name-sake creatures of these islands, even though he acknowledged that their numbers had greatly declined. "The flesh of this animal is largely employed, both fresh and salted; and a beautifully clear oil is prepared from the fat," he wrote. On Santiago Island, he and his servants camped out for a week while the rest of the *Beagle* crew searched for water. There on the coast they encountered a group of Spaniards who were fishing and salting tortoise meat, and who'd built a "hovel" up in the highlands. Darwin and his servants visited the Spaniards' camp twice. "While staying in this upper region," he wrote, "we lived entirely upon tortoise-meat: the breast-plate roasted (as the Gauchos do—*carne con cuero*), with the flesh on it, is very good; and the young tortoises make excellent soup; but the meat to my taste is indifferent."

———————

The biological oddities on the islands continued to intrigue Darwin, but his deepest convictions must have been challenged by the sheer variety of species he encountered. As science writer David Quammen wrote in his award-winning book *The Song of the Dodo: Island Biogeography in an Age of Extinction,* "God had supposedly stopped creating after the sixth day. But now, as the wider world opened to the taking of a more thorough biological inventory it seemed that God had stayed busier than anybody had dreamed."

Darwin noted, correctly, that most Galápagos animal species had no defensive adaptations whatsoever. Throughout history, many attempts have been made to define this strange phenomenon with terms such as stupid, innocent, unwary, shy, tame, tranquil. Quammen calls it not tameness, but "ecological naïveté," a trait born in an isolated world. My own term among some species is "natural indifference." Blue-footed boobies and waved albatross have performed mating rituals right before my eyes. Pelicans have brooded their newborn chicks in nests two feet away as I watched in silence. Flightless cormorants have stood before me airing their mutated appendages without a care in the world. Why? Because evolution and adaptation have stripped them of their vulnerability. Like most visitors to the islands, I also experienced the delight of what I call "natural curiosity": sea lions have peered into my mask while I was snorkeling. Tiny penguins have swum in circles around me in shallow surf. A hawk alighted on a branch above me and stared down at me with amber eyes.

Darwin found the birds of the Galápagos especially tame. "All of them are often approached sufficiently near to be killed with a switch, and sometimes, as I myself tried, with a cap or hat. A gun here is almost superfluous; for with the muzzle I pushed a hawk off the branch of a tree." On Floreana Island he watched a boy with a switch sitting by a well where finches and doves came to drink. The birds were so unwary that in a matter of minutes, Darwin wrote, the boy had collected "a little heap of them for his dinner."

Darwin was becoming a keen observer of the natural world. But how did he solve the mystery of mysteries—the evolution of life by natural selection? The fact is Darwin's role in the Galápagos persists as one of the greatest myths of the history of science. First of all, Darwin spent five weeks in the Galápagos and visited only four major islands.

Contrary to historical accounts, he was much less interested in finches than in mockingbirds, which he called "mocking-thrushes."

Discrepancies in the record of Darwin's tour and his later writings deserved scrutiny, and in the 1980s a Harvard professor and the recipient of a MacArthur Foundation "genius fellowship" turned the Darwin myth on its head. Frank J. Sulloway wrote that while in the Galápagos, Darwin had no eureka flashes of enlightenment. Nor was he convinced that all 13 species of finch collected had descended from a single ancestor. As for proof of evolution by natural selection, it would take decades before his final theory transcended his religious beliefs and his enduring doubts.

In addition, Darwin failed to identify which islands his finches came from. Luckily, three other crew members, who weren't scientists, collected and identified their own specimens: Captain FitzRoy, a sailor named Harry Fuller, and Darwin's manservant, Syms Covington. Darwin also ignored tips from Vice-Governor Lawson on Floreana Island about the variation in the tortoises' shells from island to island. Darwin saw the tortoises on only two islands and noted very little difference between them. Over the years, Darwin's failure to study the tortoises or to identify the finches would deeply haunt him.

On his return to England in 1836 he turned his collections over to the country's chief ornithologist, John Gould. Gould set to work organizing the material. But first he needed the finches from FitzRoy and crew to fill in the gaps; their specimens were far more organized than Darwin's. In his lab studies Gould identified 13 species of ground finch found nowhere else on Earth, among them finches that feed on prickly pear cactuses; finches that drink the blood of masked boobies; and carpenter finches that use twigs or cactus spines to spear insects. But Gould's most important finding was that each specimen had come from a different island. That idea titillated Darwin, who began to ponder whether the finches had descended from a common ancestor.

In 1839 Darwin published his classic *Voyage of the Beagle,* but the 500-page tome contained only 28 pages about the Galápagos. That same year he married his cousin Emma Wedgwood. The two had ten children. Those who survived tuberculosis and the other diseases of the era grew up to become bankers, botanists, astronomers, and engineers. By all accounts, it was a happy and close-knit family. Emma had studied classical piano with Frederic Chopin and often entertained. Darwin remained a quiet thinker who bred pigeons, primroses, and orchids, studied barnacles and worms, and continued chiseling away at the puzzle of how species evolve. He began reading the works of Thomas Robert Malthus, the late 18th-century economist. Darwin was struck by Malthusian ideas about overpopulation, specifically that plants, animals, and humans produce far too many offspring for available resources to support them. As a result, the weak are culled and those with favorable traits adapt, survive, and ultimately evolve into new species. Darwin had finally hit on something: the theory of evolution by natural selection.

When Darwin's favorite child, Anne, died in 1851 at the age of ten, he was devastated. He became more detached from the church and ultimately renounced Christianity; Emma feared his soul would be damned to hell. He believed that some kind of divine *energy* had created the cosmos, not a personal savior; nor could God's existence ever be proven by science. The country gentleman continued working in secret—and healing his heart—on his Shrewsbury estate. Often he'd traverse the countryside, tapping the road with his cane while deep in thought. But his quiet intellectual world would radically change when in 1858 he received a manuscript from a young British naturalist named Alfred Russel Wallace. The paper focused on the very treasure Darwin had guarded so closely for nearly 20 years—natural selection. Wallace, who came from a humble background, had done extensive fieldwork in the Amazon Basin and the Malay Archipelago. He was interested in the geographical distribution of flora, fauna, and insects, and his insights into the origin of species were quite sophisticated.

Fearing this young maverick might scoop him, Darwin panicked. He contacted Charles Lyell (whose book on geology had so upset Fitz-Roy onboard the *Beagle*), and botanist Joseph Hooker. The two scientists decided that Darwin and Wallace should announce their concept together, which they did. But it was the country gentleman Darwin, not the impoverished naturalist Wallace, who received all the glory. The next year Darwin published *On the Origin of Species,* a book that rocked the world. It declared that life is not created by divine intervention but that all life has a common descent. Species diversify from a single type by adapting to their environments. The main impetus for biological change is natural selection—meaning that species evolve along continuous lineages and that only the strongest survive. Darwin based his evidence on four themes: biogeography—the science of how species are distributed across the planet; paleontology—the geological fossil record of extinct life-forms; embryology—the development of embryos; and morphology—the shape and design of an organism's anatomy.

As for Darwin's celebrated finches, his field data were so spotty that he never mentioned them in *On the Origin of Species.* The giant tortoises are another story: Not only did he fail to collect any of the reptiles for scientific inquiry, he and his crewmates apparently ate the last of 30 tortoises on board the *Beagle* during the ship's journey to Tahiti.

Darwin continued writing and working in his garden. In 1866 he began to grow a long white beard. His hairline had greatly receded although he wore his hair long, and his bushy eyebrows protruded above sad blue eyes. No longer did he believe in a divine being or in creation by intelligent design. He wrote: "We can no longer argue that, for instance, the beautiful hinge of a bivalve shell must have been made by an intelligent being, like the hinge of a door by man. There seems to be no more design in the variability of organic beings and in the action of natural selection, than in the course which the wind blows. Everything in nature is the result of fixed laws." Nor did he believe in a God that allowed slavery, poverty, and the depth of suffering he'd witnessed among the natives of Tierra del Fuego during the *Beagle* voyage in the 1830s. Toward the end of his life, he concluded, "The mystery of the beginning of all things is insoluble to us; and I for one must be content to remain an agnostic."

For Darwin, it had taken more than two decades for the true significance of the Galápagos and their divergent species to sink in. The islands that would make him so famous were more like a postscript to his theory of evolution by natural selection, and in the first draft of his *Autobiography* he didn't even mention his trip to the Galápagos. But the islands and their distinct biology had provided a catalyst for a determined genius who refused to give up—and whose legacy would forever change the world.

CHAPTER 4

Early Years

"Conservation is a state of harmony between men and land."
—Aldo Leopold

Darwin had exposed a brave world and had shifted a deeply engrained paradigm, and it wasn't long before others arrived to explore the Galápagos Islands. Among them was Rollo Beck, an ornithologist and collector from Los Gatos, California, who had grown up working in fruit orchards. Beck, who did not believe in Darwin's theories, led a yearlong expedition to the Galápagos in 1905–06 with a team of scientists from the California Academy of Sciences in San Francisco. The turn of the 20th century marked a time when scientists worried that the unique flora and fauna of the Galápagos might soon perish, and with it the biodiversity that had made the islands so alluring.

Leverett Mills Loomis, the director of the California Academy of Sciences, believed that a research and collecting expedition to the Galápagos would put the academy on the map as a world-class institution. The odd species from the islands would jazz up its collection of mounted wildlife and create an enduring legacy. Beck was a well-known collector of bird specimens who had visited the Galápagos several times and was therefore the logical choice to lead the expedition. He and his team set out to capture as many specimens as possible, especially marine iguanas, Darwin's finches, and giant land tortoises.

About three months after they left San Francisco, Beck and his crew landed in the Galápagos where they killed doves, stole birds' nests with eggs in them, grabbed marine iguanas by the tail, and skinned and stuffed them aboard their ship. Despite their own actions, the men were shocked to see how much damage humans and invasive species had already wreaked on the islands. Beck wrote that on Española Island, "I crossed island to look at tortoise ground. Found considerable ground suitable for them but doubt if they could live now as goats eat cactus as it falls. Goat trails to every good sized cactus tree."

Beck and his teams killed nearly every tortoise they found, justifying their actions in the name of scientific research. When they returned to San Francisco in 1906, they carried in the ship's hold 266 Galápagos tortoises, most of them dead. Among them was the last tortoise from Fernandina Island. According to the California Academy of Sciences record they also collected more than 8,000 birds, 1,000 invertebrate fossils, 13,000 insects, 10,000 plants, 800 clutches of eggs, and nearly 4,000 reptiles.

G. T. Corley Smith, British Ambassador to Quito in the 1960s, would later serve 10 years as Secretary-General of the Charles Darwin Foundation, and for 12 years as editor of its newsletter, *Notícias de Galápagos*. In a 1979 retrospective, Smith wrote of Beck and his team of collectors:

> Their year-long expedition contributed much to knowledge, nothing to conservation. They collected voraciously and depleted still further the dwindling stock of wildlife. Conservation was a concept virtually unknown to their generation. Scientists simply accepted that the Galápagos fauna was doomed to extinction and that their duty to posterity was to preserve as much as they could in museums. There is no point in blaming them, any more than the settlers or the whalers; they were acting in accordance with the ideas of their times.

William Beebe was a prominent ornithologist who visited the Galá-pagos twice, in 1923 and in 1925. Beebe recorded his discoveries in a writing style that resonated with readers, even though he often boasted and exaggerated. His book, *Galápagos: World's End,* was so provocative that it became a best seller and is still widely read. He wrote: "A first walk in any new country is one of the things which makes life on this planet worth being grateful for, and in a wonderland such as this it is an event so absorbing, so replete with … new impressions that any coherent account of it is almost impossible." Like Rollo Beck, who was his friend, Beebe was a collector. "Here were cones, turrets, conches, glorious murex, chitons and rare colored cowry shells, and I found myself on my knees, picking and choosing, filling pockets and handkerchief and becoming more and more miserly and avaricious by the moment."

On the west coast of Isabela Island, near Tagus Cove, Beebe found two flightless cormorants incubating eggs on their nests. Even though his crew was filming the event, Beebe wanted the real deal. As he grabbed one of the females, "she raked my hand fore and aft with the cruel curved tip of her beak. Three times she went to the bone before I could secure her." Then one of his crewmates seized her single egg. On Isabela Island they skinned an enormous land tortoise, a species Beebe called the "patriarch of the mountain." He also captured several Galá-pagos penguins to put in the Bronx Zoo in New York. He plunked the sole live tortoise he'd taken overboard to see whether it could swim. Despite its noblest efforts to survive, the poor creature soon sank to its death.

Yet Beebe also expressed concern about some species going extinct. On South Seymour Island he was fascinated by the land iguana, which he referred to with the scientific name *Conolophus subcrestalus.* "I have eaten many scores of Mexican and South American iguanas and found them delicious, and I can readily believe some of the old voyagers who dwell on the toothsomeness of *Conolophus*. But every one we collected was too precious to sample."

Beebe's portrait of the Galápagos was like a beacon to Europeans who wanted to flee their war-torn nations. They migrated to this far-out world, beginning with a group of Norwegians in 1926. They created the town of Puerto Ayora and started a short-lived fishing industry. In 1928 a German physician named Friedrich Ritter and his companion, Dora Strauch, moved to Floreana Island. In 1932 Margaret and Heinz Wittmer also moved to Floreana seeking a healthier life for their disabled son, and became successful farmers. Floreana also attracted an eccentric Austrian woman who fancied herself the empress of the island. In 1933 she arrived with two lovers, Robert Philipson and Rudolf Lorenz, to build a hotel for the wealthy yachtsmen who were frequent visitors to the island. The "Baroness von Wagner de Bosquet," as she was known, was fond of carrying a gun and a whip. Despite her bullying, workers hired to build the hotel constructed nothing more than a corrugated metal shack. In time, the fickle and controlling baroness chose Philipson as her lover, rejecting Lorenz, who fled in terror and died when he became shipwrecked on the island of Marchena at the northwest end of the archipelago; his mummified body, stiff in the equatorial sun, was found some time later.

Fearing an attack on the Panama Canal, Franklin D. Roosevelt began to negotiate with Quito in 1938 to set up an air base in the Galápagos. Agreement wasn't reached until 1942. The U.S. Army Air Corps was allowed to establish a base on Baltra Island. The base was huge, staffed by 2,500 U.S. and Ecuadorian servicemen and staff. The soldiers called the island the "Rock," like Alcatraz, because there was no boat on which to leave the barren and desolate landscape. Construction of the airport, landing strip, barracks, cafeteria, bar, hospital and other buildings took a serious toll on the land iguanas for which the Rock had been home. Rumors flew that bored soldiers had picked off the reptiles one by one as target practice. In 1942 the Secretary of War ordered that the base commanders "take appropriate action to prevent any unnecessary molestation of the wild life in the Galápagos Archipelago and to prohibit the introduction of domestic animals that may prey on the native fauna." Both the U.S. State Department and the Smithsonian Institution joined in monitoring the iguanas. Whatever happened to these four-foot-long reptiles is still unclear. By the time the American base was turned over to Ecuadorian authorities in 1946, the rare iguanas had disappeared. (A colony of Baltra iguanas removed to nearby North Seymour Island in 1932, however, survived; in the 1980s the reptiles entered a captive-breeding program at the Charles Darwin Research Station, where they were raised and repatriated to Baltra in 1991. Today, about 400 land iguanas live on the island.)

In 1932 Ecuador had enacted laws to protect the islands, and in 1936 the entire archipelago was declared a national park. But it took more than 20 years for the real work to begin. A Briton, Sir Julian Huxley, chair of the International Zoological Congress, pushed hard in Europe to create the Charles Darwin Research Station (CDRS), which opened in 1964 on Academy Bay in Puerto Ayora. The Charles Darwin Foundation (CDF) had been organized in 1959 to oversee the research station and advise the new park. The mission of the CDF today is to become the world's leading research institution, to conserve the biological diversity and natural resources of the Galápagos, and to create a sustainable society that understands the value of this bioregion and is committed to protecting it.

When the CDRS opened in 1964, it was headed by a visionary named Victor Van Straelen, who served as the foundation's first president. By then, the concept of conservation was catching on, and 3,000 of the world's most distinguished zoologists urged the International Zoological

Congress to save the Galápagos. In 1965 a captive-breeding and incubation program was started at the research station with tortoise eggs from Pinzón Island, where black rats had killed every single hatchling over the preceding 50 years. Under the direction of ornithologist and research station director Peter Kramer, a committee drafted a vital "Master Plan for the Galápagos National Park" as the foundation for future governance. When Kramer left his post in 1974, tortoise expert Craig Mac-Farland took over. Soon more than 500 scientific missions would travel to the Galápagos—and, not surprisingly, hundreds of tourists longed to visit the islands. But MacFarland knew that if this were to happen, the impact of so many visitors would need constant monitoring, and park wardens and naturalists would need proper training to guide them.

In 1978 the United Nations Educational, Scientific, and Cultural Organization (UNESCO) designated the Galápagos National Park as one of the very first World Heritage sites. The goals of UNESCO were: to prevent invasive species from entering through educational awareness and a strict quarantine system; to eradicate invasive species that had already gained a foothold on the islands; and to promote sustainability over time. The UNESCO World Heritage Centre would work closely with the CDRS and the Galápagos National Park. In 1984 UNESCO gave the Galápagos even more protection as an International Biosphere Reserve "to promote and demonstrate a balanced relationship between humans and the biosphere."

Scientists continued to flock to the CDRS from all over the world to study the strange life-forms of the islands. But its facilities were inadequate to house them. In the early 1960s tourism had also gained a small foothold in the Galápagos. Among the earliest tour operators were the Angermeyer family, who had fled Germany before World War II; the Wittmers, who had settled on Floreana; and an American sailor and expatriate named Forrest Nelson. In 1960 the robust Nelson built the first hotel and restaurant on the islands—the famous and now-closed

Hotel Galápagos on the rocky shoreline of Academy Bay. It was within walking distance of the research station, just a few minutes down a sandy trail. Nelson's hotel made it easier for visitors, and soon he was organizing the first ecotourism trips in the islands. But getting to the Galápagos back then was absolutely hellish. Flights from the mainland, offered through the Ecuadorian military, were unscheduled and often nonexistent. Supply barges served the islands only a few times a year.

Jack Nelson, Forrest's son, remembers those days well.

In a word, we were isolated. Isolation made us both interdependent and self-reliant. Above all, it was the patchy transportation that enforced our isolation. Small cargo ships were often our only option. My first voyage from the Galápagos to Guayaquil was in 1967 on the *Santa Cruz*. She had the beautiful riveted steel construction and style typical of coastwise traders of 1910. She looked like something from a Bogart movie. When a piston broke in the main engine, the mechanics simply removed the junk parts and the ship continued to run with the rollicking syncopation of the missing beat. This is the way of life in these little struggling republics; you wing it with the tools and materials at hand. All these crummy vessels carried live cattle on the return trip from Galápagos to the mainland. Cows, goats, and pigs. Yes, the legendary South American Cattle Boat, four days to seamy Guayaquil, with over forty cows on deck, a heap of pigs under the foredeck, and four-hundred-fifty goats in the hold. Stink? The word is inadequate. Even the air tasted vile.

Pigs resolve minor quarrels with a screeching tusked menace, but as our shipmates they settled into a minimum of resistance; an equilibrium of discomforts. Pigs, like humans, adapt to almost any awful situation. But whoever has been seasick must cede the glory of suffering to a cow tied by the horns at a ship's rail; imagine being seasick in four stomachs. In the open hold were hundreds of goats. They bleated and shrieked and reeked as imps in a dungeon, the buckest billies striving for supremacy of a higher perch in the ship's frames, the rest roiling like a furry ant heap. We ate whatever died.

———————————

Meanwhile, down the road from the Hotel Galápagos (where I met with Jack Nelson many times over the years), scientists and park rangers in the early 1970s were busy monitoring endangered species. The giant land tortoises were especially vulnerable. The park, in conjunction with the CDRS, as mentioned, had begun a captive-breeding program to incubate the eggs, raise the hatchlings, and repatriate the tortoises to their native islands. But the human face is important here, and that's where the story becomes interesting. What I mean to say is that my understanding of the Galápagos went through a complete metamorphosis when I unexpectedly met the last of a breed—a native Galapagueño and 66-year-old pioneer—one of the most brilliant naturalists in the park's history.

———————————

Oswaldo Chapi's once black hair is almost completely gray but still as thick as a horse's tail. Square bifocals rest on his nose just above a scar, slipping as he walks. A generous belly pokes out above his khaki shorts, but his muscular legs bespeak a lifetime of working outdoors. On a drizzly July morning on Santa Cruz Island, Chapi leads a group of visitors and me through a canopy of candelabra cactus to the Galápagos National Park, home to some captive giant tortoises. Darwin's finches hop from limb to limb in nearby trees, nabbing insects as they go, and a yellow warbler sings sweetly from a tangle of brush. Chapi, a retired Galápagos National Park ranger, still leads group tours from time to time, and soon the reason for this becomes clear. This native naturalist knows more about tortoises and their behavior than just about anyone, and scientists from around the world contact him for his impeccable knowledge.

Chapi, who was born on a ranch in the highlands, never completed grade school because there was no school up there at the time. Yet he knows the name of every plant, animal, and insect, in Spanish, English, and Latin. There's something to be said for the knowledge of the self-taught. Christy Gallardo Nelson, Forrest's daughter and Jack's older sister, has known Chapi since she was a teenager. Once, at her home, which is full of her own exquisite art, she showed me an old black-and-white photograph. In the picture Chapi, a slim young man, is leaning against a tree, with Christy and her son Jason alongside. "Look at him," she says. "He's one of the people I admire most here. If you blindfolded him and spun him around, then asked him to parachute out of an airplane still blindfolded, when he landed, he'd be able to tell you, 'That's north, and that's south, and just beyond that hill mushrooms are growing next to ferns.' " She's absolutely right.

This chapter of the story begins in the early 1970s when, as one of the first rangers for the Galápagos National Park, Chapi and other young men were sent out to scour the volcanic islands for signs of giant land tortoises, their hatchlings, and even their eggs. In the few decades that humans had lived on the islands, the toll on the wildlife had already been devastating. One of the biggest menaces throughout the archipelago was invasive species. Feral goats, burros, pigs, dogs, and cats brought to the islands by pirates, whalers, or colonists preyed on the wildlife and decimated habitats. The tortoises in the wild had been especially vulnerable.

Now, as we stop at a pen that serves as home to two large males, Chapi squats down to greet them. "Hello, how are you?" he asks softly. "How have you been, my friends?" Slowly, their alien-looking heads poke out of drab green carapaces, and their eyes cock upward in his direction. Maybe it's his voice or his scent, but there's a connection here that's hard to fathom: They're like family. Chapi knows that these magnificent reptiles he once saved from extinction will outlive him, his children, and possibly his grand-children. As I stand among the group of visitors this misty day I'm truly moved. This is my third visit to the park to observe these giants—but it's the first time I've ever seen a naturalist guide treat them as *equals,* not icons.

We cross a small bridge leading to another pen. There, in a shallow pool, sits the very symbol of the Galápagos: Jorge Solitario, Lonesome George, the last of his breed from the island of Pinta. Chapi knows this old reptile well. In 1972 he and another ranger carried the behemoth off Pinta Island in hopes he would breed with a female with similar genes from a neighboring island. "We've tried everything," Chapi says, his bushy eyebrows furrowing into a v, "but he will *not* mate." The research station even flew in a young researcher to collect sperm for artificial insemination, but George would not give in to her gentle coaxing.

When Chapi recalls his younger days, a hint of melancholy surfaces in his dark brown eyes. Back then he could run up and down volcanic slopes without losing his breath and survive long hours under the equatorial sun without water. Given the topography of the islands, I'm curious to know how he and the rangers got the huge tortoises down the mountain to the awaiting boats. Some of the animals weigh as much as 200 pounds. "We didn't have the equipment to carry them, so we tied them to our chests or backs with cord," he explains, patting his chest with his palm. "They had never been off the ground before, so for them it was like flying. They were so scared they shit all over us. By the time we got down the mountain we had to jump into the ocean to bathe."

Like his old friend Lonesome George, Chapi may be the last of a breed as the Galápagos bend to pressures imposed by oil spills, widespread corruption, lack of enforcement of the laws, greed, lack of education, and skewed policies where authorities often look the other way. All of these factors have contributed to the fragmented ecosystems, extinct and endangered species, and dead zones of the Galápagos today. Now there's a new trend in the islands: a surge in fundamentalist groups who believe the world is only 6,000 years old.

A day earlier I had been on a day trip to Bartolomé Island with a naturalist guide who was studying the Bible of the Jehovah's Witnesses and who told me he did not believe in evolution. When I tell Chapi about this and mention that I've met other creationists in the Galápagos, he visibly bristles.

"They've been converted by the Bible," he says. "*Están fuera de la realidad*—They're outside of reality." Then he explains why. "The woodpecker finch uses the spine of a cactus to pick insects out of wood. It's a kind of intelligence that gets passed down by its ancestors. It's environmental adaptation. The flightless cormorant builds its nest with marine algae because it's so dry there aren't any leaves or other organic materials to build with. The vampire finch drinks the blood of the masked booby when there's no water or anything to eat. It's a question of evolution changing organisms so they can survive."

Obviously Chapi knows natural science. "I love nature," he concludes. "This is my life, to learn how organisms eat, sleep, hear, see, smell, survive—everything."

In April 2008 I run into Chapi on Avenida Charles Darwin. He is walking back into town from the research station with a group of young American students he had just taken to the tortoise pens. Chapi hugs me; he seems happy to see me. Maybe it's because I always ask him so many questions. He tells me that the students he's leading today are really bright. "They want to know about everything." Then he says that he wrote a 200-page book about the Galápagos but someone stole his computer. Chapi purses his full lips and stares down at the cobblestone street. "Now I need to try to remember everything." The loss is tragic, and so is the loss of his father's farm in the highlands, where he grew up. He can't afford to run it. "I'm trying to sell the *finca* [farm] with my brother, but there are too many permits required and they're very expensive."

"What a shame," I say.

"*Sí, mi hija,*" he replies wistfully. "Yes, my daughter." But now Chapi has to catch up with his students. Before parting I tell him how much I respect him. "You're the perfect symbol of everything the Galápagos stands for."

"*Era,*" he replies, using the past tense. "Was." I'm not sure whether he's referring to himself or to the archipelago.

A few months later I return to visit Lonesome George. He's acting weird, poking his head under the saddle of one of his female companions as if to nip her. His behavior confirms mounting reports to the research station and the park that George isn't his normal indifferent self. Then the news hits like a bomb. The 90-year-old virgin has finally discovered sex and has mated with *both* of the females from Wolf Island. His caretakers are stunned to discover three nests, each with several eggs, in his enclosure. I return to visit George, expecting a media frenzy, but the only people there are a couple of geologists who've been working on Fernandina Island. The husband-and-wife team is sitting on the cement wall above the enclosure, silently watching him.

"Amazing, isn't it?" I say.

"Yeah," replies the wife. "We came to congratulate him."

In the months that followed, tortoise researchers, like expectant mothers, monitored the eggs around the clock. But in the end, not one of the eggs proved fertile. When Lonesome George dies so will his lineage—one more extinct species in paradise.

CHAPTER 5

Puerto Ayora

*"Find your place on the planet.
Dig in, and take responsibility from there."*

—Gary Snyder

It rained a lot here in 2008, more than anyone could remember since the last El Niño in 1997–98, when the climate shifted so radically that endemic species such as the marine iguana and Galápagos penguin nearly perished. This year, throughout Puerto Ayora, roofs turned into sieves; homes and stores flooded. Green mold dusted the inner walls of my house, my clothes, even my books. At dusk mosquitoes descended in clouds, flying ants swarmed streetlights, and centipedes longer than pencils wriggled out of their burrows. Then locusts descended on red and orange wings and covered the streets like 1970s-style shag carpets. Yet no one knew what to call this strange new phenomenon. For one thing, the ocean currents flowing south from Central America weren't warm enough to classify this as an El Niño event. Scientists and residents alike scratched their heads, saying, "weird."

Like most boomtowns, Puerto Ayora doesn't have the infrastructure to withstand such aberrant weather, nor does it have the resources to support the 18,000 or so residents who've pushed the town to the boundaries of the Galápagos National Park. Basic services are scarce. Take water, for example. The water on Santa Cruz Island is too brackish to drink and must be purified, or water must be shipped from the mainland in plastic bottles—by the millions. A faulty septic system sends effluent into the bay where children swim, yet health care in the town is completely inadequate. During the eight months I lived in Puerto Ayora, residents suffered from a variety of ailments: persistent coughs, skin rashes, and vaginal infections. On several occasions, intestinal bacteria sent scores of children to the clinic and the *farmacias* ran out of bottled rehydrating solution. Parents who could afford to, flew their children to hospitals in Guayaquil on the Ecuadorian coast.

In Puerto Ayora, electricity runs on about 57 cycles instead of 60, according to residents who have lived here for decades. Without a hefty transformer a computer can crash and burn, literally. Most houses don't have a hot water heater; dishes are washed in cold, unpurified water. For a warm shower, I have to rely on electricity. Here's how it works: A jumble of electrical wires connect from a device on the showerhead to a wall switch. To heat the water one must flip the switch. Galapagueños call this a "Frankenstein shower"—a perfect eponym. Each time I flip the switch blue sparks fly from the wires, sending shock waves through my hand and standing my hair on end. This is not an exaggeration. The water never really gets hot, either; barely tepid. To rinse all the soap away, I pour a bottle of purified water—or filtered rainwater—over my head. On small islands one learns to adapt.

One evening as I eat my dinner up on my tiny *terraza* I am stunned when a blackout engulfs the entire town. Puerto Ayora has vanished, absorbed into the night in a second. Soon, candles begin to appear in windows. Tourists wander down Avenida Charles Darwin giggling like children as dogs yap from the shadows, reacting as if a lunar eclipse has blotted out the moon. In the Southern Hemisphere, the Big Dipper hangs upside down. Mars is a ruby in a sea of ink. For a moment I recall that *this* is what Puerto Ayora looked like when I first visited nearly 20 years before: a sleepy fishing village on the edge of the bay with only a few thousand residents who depended on candles, lanterns, or electricity rationed from gasoline generators. But my memories shatter when a taxi passes on the roadway below, blaring Ecuadorian rap music. The bass is so loud it vibrates through the soles of my feet.

When the Charles Darwin Research Station (CDRS) opened in 1964 to support the Galápagos National Park, Puerto Ayora was home to only a few hundred people. Today, the growing demands of tourism have transformed the town into what many describe as "Disneylandia." Tourist shops stocked with trinkets and overpriced T-shirts dominate the main street. The most popular shirts and caps proclaim: "I love boobies." One T-shirt shows a giant land tortoise copping to his mate, "So I was unfaithful, but that was 150 years ago." At Bar El Bongo, whose western wall abuts the Seventh-day Adventist church, visitors can swig "body shots" off the semi-clothed torso of anyone willing to become a human bar. A bridal reception here in 2008 featured a male stripper whom one attendant described as "really hot."

Then there's the new fountain whose centerpiece is a Dutch-style windmill that spills water over ceramic tiles covered with algae. In a town where dengue fever arrived only a few years ago, people wonder why such an obvious beacon for mosquitoes was built in the first place, and why the money wasn't spent on environmental education—or health. Pelican Bay lies just across the street from the fountain. In 1990, when the main road was made of volcanic cinder, I stood here watching traditional fishermen build sleek wooden boats. Today, the site is a concrete slab abutting the bay where men in tall rubber boots clean their daily catch. As they toss the heads and tails of grouper and other fish to the pelicans that congregate like a pack of dogs, the birds rise up in a mosaic of wings, their quivering gullets waiting to be filled. The scene is a big hit with camera-toting tourists, especially when a baby sea lion hops onto the dock, barks at the pelicans, and steals the show.

Meanwhile, out on the street, taxis pass nonstop from early morning until late at night, running on subsidized fuel in this petroleum-rich nation where premium gasoline (in July 2008) cost only $1.43 a gallon. At the same time, in Miami, premium gas had surpassed $4 a gallon. Puerto Ayora has at least 400 taxis, little white pickups where drivers slam on the brakes when a marine iguana crosses the road, or when an engorged pelican waddles across to the fountain to splash around and preen. Like many of the locals, I've shooed birds and marine iguanas back to the bay, flapping my skirt at them as if I were herding geese.

Wildlife in the Galápagos knows no boundaries between the natural world and that of humans.

Welcome to the Galápagos, where a quart of pilsner costs a mere $2. But order a steak dinner or sushi at one of the nicer restaurants and you could pay up to $18, plus tax and service charge. Puerto Ayora is fairly tidy along the main tourist street, where big blue bins urge people to recycle. A few blocks north a different world emerges. Gone are the recycling bins, replaced by piles of litter. Dogs run loose without collars or tags. Cats stalk yellow warblers and Darwin's finches. In some yards chickens yank at tethers tied to their legs, even though chickens (and goats) are illegal within the town's limits. In every neighborhood, rods of rebar poke up from the roofs of cinder block homes—testaments to the unfinished dreams of a new prosperity. In some sectors, broken down buses and rotting boats lie cradled in weeds. Bars display posters of busty babes alongside pictures of Jesus. A few have TVs, most of watch are permanently tuned to the national pastime, soccer. On game day young fans ride on bus tops, shouting slogans through megaphones.

Many of these young people—and young couples with children—still reside with their parents because there's nowhere else to live.

I know because Puerto Ayora is a small island town. Some days a 15-minute walk to the main market might take me an hour because shopkeepers along the way love to come out and chat. These exchanges delight me and provide a sense of community where none really exists. I learn a lot about the townspeople, their families, and their concerns about the future. I have also become more adept at picking up cultural nuances and some interesting slang. Take the word *"pluto,"* for instance. It means "drunk." A man who has imbibed too much *está pluto* (he's drunk). A woman *está pluta*. Galapagueños and other Ecuadorians were using this trope long before the smallest "planet" got dissed and the 2007 "Science Word of the Year" became "Plutoed." Then there's the "gringo fish," local slang for the Creole fish, a grouper whose red belly resembles all those sunburned tourists.

If not for tourism, Puerto Ayora might have remained a quiet fishing village where pioneers didn't need money. Instead, they would barter fish for fresh produce from the fertile highlands where oranges, papayas, and avocados the size of grapefruits grow. At one time there were so many avocados that farmers fed them to their cattle. Back in the 1960s and 1970s most visitors wore backpacks and arrived either by sailboat or on stinky cattle barges from mainland Ecuador. These free spirits cared deeply about the Galápagos and longed to learn every detail about the islands' natural history. They asked questions like: Why do penguins and flamingos coexist at the Equator? Why are the booby's feet blue—or red? How long can a marine iguana hold its breath under water? Why are reptiles the dominant species on these islands and not mammals? Why don't whitetip reef sharks eat me when I swim close enough to touch them?

Naturalist guides for the Galápagos National Park agree that most tourists today visit the islands for very different reasons. "In the past, passengers wanted to know every single detail," says Ivonne Torres, co-owner of the travel agency Unique Ecuador and a naturalist guide for the national park, whose English is impeccable. "Now they're more interested in having a good time. Yes, they want to see nature. They want to see the most they can. But very seldom do you get challenging questions."

Torres is a Rubenesque woman with tinges of red in her jet-black hair. She's charismatic and popular among tourists, and her knowledge of these islands is extraordinary. Like other guides who work on the larger boats, she laments the stark absence of engagement. "Passengers have *devolved* instead of evolved," she says, adding that the Galápagos are just one more stop on a combined tour of the Amazon rain forest or the ruins at Machu Picchu. She's right. It's common to see tourists walking around town, their khaki vests plastered with patches from every exotic place imaginable.

Most professional guides agree. Patricia Stucki, an athletic woman who grew up in Switzerland, is one of the top naturalist guides in the Galápagos. Before coming to the islands, Stucki lived in Costa Rica, where she guided visitors up active volcanoes on horseback or took them diving around islands off the coast of Honduras. She moved to the Galápagos in 1996 and was soon hired as a diving guide in a male-dominated field. She has worked as a naturalist guide and dive master for the park, and a dive instructor for the Galápagos Marine Reserve (GMR), the world's second largest marine reserve, established in 1998. Stucki concurs that the caliber of tourists—and their motivation for coming—has radically changed. "Most people," she says, "care less about pristine nature than visiting an exotic destination that's on *the list*. You have to check it off. There is the Chinese Wall, check it off. Having children, check it off.

Going to Galápagos, check it off." Stucki's hair is bound up with a single chopstick. As she removes the stick and shakes her head, blond waves cascade all the way down to her waist. Here's a highly educated woman who is fluent in five languages—her native Swiss German dialect, plus German, French, Spanish, and English. She's frustrated, too, because most visitors now ask very few questions. Nor do they listen. "Let's say I've been talking about the marine iguana's diet for ten minutes and I finish, and then somebody asks, 'And what does the marine iguana feed on?' Do you explain it all over again? It's something that eats you up through the years, eats up your energy. You have to be patient."

But Stucki does make a difference. Part of her job as a guide is to help people realize what's "already inside them," as she says, to help push them beyond self-doubt and fear: "To live." Once, she received a card from a passenger who wrote: "My mother was pretty convinced that she wouldn't do any snorkeling on this trip, and you got her out there and now she cannot wait to go snorkeling again." Stucki also encourages tourists to slow down and open their senses—to see and hear and smell—and to pay attention to the details when they return home.

I ask her what three things she'd like to see change in the Galápagos.

The first, she says, is to stop shark finning. "It hurts me when I see sharks without fins when I'm diving. They [fishermen] just slaughter them." The second is "to eliminate corruption." The third is "to set limits on tourism" and regulate how it's organized.

"Is environmental education adequate here?"

"It's like a few drops on a hot rock; they just evaporate. It helps to plant a certain consciousness in people's minds, to give them reasons to be proud to live here. [But] education is so basic that it's hard to open people to the environment. We have to start with children, and I think we are. Some of the children are [now] educating their parents."

Visiting the Galápagos as a tourist is one thing, but living here is radically different. Paradise is what you make of it. There's heaven and hell in every second. For one thing, Puerto Ayora has no access to the sea. It's not a place where I can walk out my door, spread a towel in the sand, and body surf in crystalline waves. The town is built around Academy Bay, where boats and barges leak fuel into the water, effluent seeps in, and fishermen clean their catch right there, attracting sharks. That's why I've found sanctuaries a bike ride away from this eclectic boomtown. My favorite place is Tortuga Bay, so named because sea turtles still nest there in the fine white sand. On one part of the beach the waves are big enough to surf. True, at times, the throng of visitors is overwhelming, especially on weekends, but this is the ocean—not a polluted bay. It's a place of solace where I can walk barefoot beneath a big straw hat, swim with marine iguanas, or stand at dusk to study the line where sky and sea dissolve into one.

To get to the beach from town involves a 15-minute bike ride, followed by a 30-minute walk on a paved trail down to the shore. It's December, the tail-end of the dry season when leafless branches claw at the sky and seedpods chatter on the breeze. Fallen trees lie everywhere like heaps of bones, but this trail is no dead zone; it's a wonderland where lava lizards do push-ups and mockingbirds eye them for lunch, where the sweet trill of warblers floats through the trees, and Galápagos blue butterflies hover over flowers. They remind me of my childhood in northern rural Japan, where in spring I'd race down hills with a net, catch a butterfly, examine the dust on its wings, and then release it like an angel to its home in the sky. Even the woody scents along this trail seem familiar. I know them from the Sonoran Desert of southern Arizona: Here they are, stowaway trees from the Americas: mesquite, palo verde, and ironwood, all hearty members of the legume family. Candelabra cactus grows along the trail, its columnar arms reaching skyward, much like the organ pipe cactus. Here, too, the prickly pear cactus has morphed into a tree about 40 feet tall, perhaps to escape the reach of the hungry tortoises that once flourished on this island and feasted on its fleshy pads.

The prickly pear holds a special place in the ecological balance of the Galápagos. It's known as a keystone species because so many creatures depend on it for survival: Mockingbirds and finches feed on the nectar, and finches build their nests in the cactus soon after the rains come. When the blossoms open, black carpenter bees pollinate the waxy yellow flowers. This is a perfect example of commensalism: Each species depends on the other for survival. Few insect pollinators live in the Galápagos. Studies have shown that some plants that established themselves early on also developed the ability to self-pollinate. Most flowers in the Galápagos are either white or yellow. With so few pollinators, plants with large showy flowers, like red or purple, can't survive here.

One of my favorite flowers comes from the *muyuyo* tree, a member of the borage family. The trumpet-shape blossoms are yellow and dangle in honey-sweet clusters. Their nectar attracts carpenter bees and Galápagos sulfur butterflies. Carpenter bees are especially crafty at what's called nectar robbing. Instead of entering the flower, the bee cuts a slit at the base of the corolla, avoids pollination, and draws out the nectar instead. In the rainy season when the winds pick up, muyuyo blossoms blanket the trail to Tortuga Bay, as though a wedding procession has just passed down to the sea.

At trail's end I take off my sandals and start walking in soft white sand. A few surfers are out on Playa Brava, so named because of the undertow and the rocks. I sit in the sand near a patch of red and green carpetweed and look around. A few marine iguanas have just emerged from the surf. As they cross the beach in single file, their tails and claws leave etchings in the sand. They look like they've just stepped out of a cartoon.

The heat is stifling, so I remove my bright blue sarong and enter the sea. The water feels just right, and I float, watching pelicans ply the skies above Santa Fe, an uninhabited island to the south where I once snorkeled. The perfection of light, the bulge of the Earth, the rhythm of the waves are distant memories locked in my cells.

In the afternoon I walk down the beach to a rocky outcropping, then follow a well-marked trail up a shallow bluff. Marine iguanas appear everywhere, their torsos aligned with the sun for warmth. My presence is of no consequence whatsoever to these creatures; in a few places I must step over them on the trail. Then something catches my eye: A large male is nodding his head as he moves toward another. It's mating season, and staking out one's territory is key to survival. The iguanas move stealthily toward each other, eyes locked in a primeval ritual too old to fathom. Then the larger male rises up and charges across the trail. The attack is quick and definitive. I stand there, stunned, as the two thrash around, biting at each others' backs and tails. Sand flies as these spiny reptiles wrestle for at least five minutes. Then suddenly they stop and stand motionless—snout to snout—until the loser backs off one scaly step at a time. The alpha male has reclaimed his territory in a thicket of giant prickly pear cactuses. I feel as though I've just stepped into the Mesozoic.

Back in town I rinse off the sand in my Frankenstein shower then walk down to a restaurant called La Garrapata. It's popular among tourists, especially those traveling on cruise ships, who visit Puerto Ayora briefly to shop for souvenirs and observe the giant tortoises at the national park headquarters. These ships are known as "floating hotels" because passengers stay in town only a few hours before motoring back to their ships in *pangas* (dinghies). Because many tourists don't stay overnight, local residents complain that big tourism contributes little to the town's economic well-being.

But it's tourism that has driven the economy for the last few decades, and it's tourism that remains one of the central threats to the towns, their residents, and the ecological integrity of the islands. Commercial development appears everywhere: hotels, restaurants, and shops to support the explosion in tourism. In 2007, about 174,000 tourists visited the Galápagos, an increase of about 30,000 over the previous year. Officials say the number of visitors in 2009 could reach 200,000, even though many ships and tour boats are already booked for the next few years. (A new tourism model, due out in mid-2009, might try to limit visitors by raising the park entrance fee for foreigners from $100 to $300 per person.) As for the fares for tours and other services, it's anyone's guess.

———————

Graham Watkins, a stout Welshman with ruddy cheeks, can put things into perspective more clearly than just about anyone. I first met Watkins when he was the executive director of the Charles Darwin Foundation (CDF). His spacious home on Academy Bay is a testament to his character—and his work. The Welsh flag, emblazoned with a large red dragon, snaps in the wind just off his front porch. The interior of his home is like a museum, full of items from Guyana, where Watkins directed the Iwokrama International Centre for Rain Forest Conservation and Development. Shields and spears hang on the walls. A large coffee table supports an impressive menagerie of wildlife from the tropical rain forest, each piece expertly carved.

On a bright Tuesday morning I slide open the wooden gate to my house and walk down Avenida Charles Darwin past an endless stream of tourists to meet Watkins at his office at the CDRS. The road passes an overgrowth of shrubs and tall prickly pear cactuses. It's only 8:30, but the park is hotter than a griddle. At the research station sun-bronzed workers wear T-shirts on their heads as they move furniture around. Others repair a crumbling wall made of basalt boulders. New air-conditioning units sit on a walkway near a solar-powered building, and just offshore two satellite dishes point in opposite directions. The bay is full of boats: big tour boats, cargo barges, sailboats with flags from all over the world, and a tuna boat from Panama that park officials seized for fishing illegally in the protected marine reserve. The ship is enormous—about 250 feet long—with a helicopter parked on top. Two oversize binoculars are mounted on deck near high-tech radar devices, several fishing skiffs, and an illegal purse seine piled as high as a house.

At 9 a.m. Watkins invites me into his office, a spacious room with a broad view of the bay. Bookshelves overflow with reports, technical papers, and hundreds of books. A cup decorated with an image of Lonesome George sits on his desk, the tab of a tea bag dangling over the side. Watkins first came to the Galápagos to work as a naturalist guide in 1987 with a friend, Pete Oxford (the guide whom I would later join on a strenuous hike up Volcán Alcedo, where giant tortoises live in the wild). The first time Watkins saw the islands he immediately fell in love. "It was a magical place," he recalls, swiveling in his chair to gaze out over the bay. One of his favorite memories is the time he watched a sea lion chasing a penguin. "Both of them were porpoising, going around and around in a pool for about 15 minutes," he says, his smile revealing a space between his two front teeth. "I'm not sure the penguin was having fun, but the sea lion was. Things like that made me rethink what I was doing and where I wanted to be."

Watkins wanted to be in the Galápagos, and in 1989 he left to pursue a Ph.D. in conservation with the goal of returning to the islands. A few years later he applied to become executive director of the CDF and was turned down. But he didn't give up. The dogged Welshman kept applying, and on his fourth try he was finally hired. It was 2005 and relations between the foundation, the park, and the central government were strained to the limit. Since 2002 the Galápagos National Park had gone through 14 directors, and Ecuador itself had gone through 7 presidents since 1996. Why? Political instability on the mainland, widespread corruption, violence in the fishing sector, a stampede of colonists, and tourism run amok. Transparency International, a German-based watchdog group, has ranked Ecuador one of the most politically corrupt countries in the world.

Watkins had inherited a nightmare. When President Gustavo Noboa Bejarano left office, Lúcio Gutiérrez, a former army colonel, took over. President Gutiérrez appointed Edgar Isch as minister of the environment. Isch was the leader of the Marxist school teacher's labor union, Movimiento Popular Democrático (MPD), in Quito. Isch had never been to the Galápagos, nor did he have much knowledge of the archipelago's environment. Isch immediately dismissed the director of the Galápagos National Park, Eliécer Cruz, who was highly respected for his accomplishments throughout the islands. (Cruz would later become governor of Galápagos Province.) When the newly appointed park director arrived in Puerto Ayora, he and the head of the CDF rode around town in a motorcade as the crowds shouted: "*Now the park is ours!*"

Watkins knew what he was up against, and in the years to follow, the Galápagos became more politicized than ever. The problems he'd inherited were huge: a lack of education, especially environmental education among Galapágueños; few models for sustainability within an island culture; widespread corruption; failure to regulate and control the way businesses are built and run; political patronage in hiring; almost no vocational education; the inability to build social capital (there are no credit or community banking systems in the Galápagos); and, finally, no sense of community, hence *no identity with place.*

"You have a frontier mentality where you've got people that came here in order to make money," says Watkins. Tourism had grown so rapidly that it created a domino effect. Local businesses sprang up almost overnight to support tourism. A flood of immigrants doubled the human population on the islands within 15 years. The new colonists overfished the protected marine reserve and introduced a rash of invasive species, just as tourists and cargo ships had done in the past. Today, more non-native species exist on the islands than native ones. Add to this the amount of fuel needed to transport all those immigrants, tourists, and cargo, and you have a recipe for disaster. In a 2001 accident, a fuel tanker called the *Jessica* ran aground, spilling about 240,000 gallons of fuel in the pristine waters. The fuel was destined to a huge tourist ship awaiting disembarkation. It was an enormous wake-up call.

Watkins quickly assessed what needed to be done to protect the islands. First, he worked to improve the relationship between the CDF and a deeply divided park service—and community. He looked at ways to improve funding and to work closely with local, national, and international organizations. He researched ways to integrate biology, geography, social sciences, and the economy, and to ground CDF policies in reality. The vision made sense, but *reality* in the Galápagos is a hard concept to grasp. Things change literally every day at the whim of whomever is in power, and solving the social, cultural, political, and economic problems on these islands has been akin to the task of Sisyphus, a mythical figure condemned for eternity to push a boulder uphill, only to watch it roll down again.

At the root of the problem is an unregulated and uncontrolled economy based on tourism, which has increased by an average of 14 percent a year. According to *Galápagos at Risk: A Socioeconomic Analysis,*

Watkins co-authored in 2007 with his colleague Felipe Cruz, a Gala-pagueño and Fulbright scholar, Galápagos tourism generates a total value of $418 million per year. An estimated $63 million of that enters the local economy. About one-fourth of that figure goes to the captains, crew, and naturalist guides on the big cruise ships and tour boats and the merchants who supply them. Another 25 percent is snapped up by the airline companies and travel agencies on the mainland and abroad. Just look up "Galápagos" on the Internet.

In contrast to the $418 million a year from tourism, fishing represents less than 4 percent of the total income-producing activity in the Galápagos. Why? Because in the 1990s *pescadores* (fishermen)— who were well organized politically—depleted the very resources they depended on for survival. With the collapse of the sea cucumber fishery, the pescadores demanded, and obtained, permits under the Special Law for Galápagos, passed in 1998, to become tour operators, shuttling passengers around on day trips without sufficient training or language skills. When fishing died in 2005, tourism shot up by almost 50 percent. "They're absolutely related," says Watkins.

At the same time, government instability on the mainland had weakened the leadership in Galápagos Province. Multinational investors began taking business away from local operators as the demand for public services and jobs escalated and migrants demanded better wages and living standards. Watkins says part of the problem is a shift in the market. "If you look at tourism in most parts of the world it starts small and it starts growing, bringing in investments, bringing in human resources, which is the same thing that's happening in Galápagos." As a result, the type of people who come here starts changing as well. Watkins is talking about sophisticated developers and investors, many of them foreigners. As the towns grow, a different kind of tourism evolves as well. People looking for lodging don't want a $40 room with a ceiling fan; they want a $150 room with air-conditioning—and maybe with a day of sportfishing (which is illegal).

By way of explanation, Watkins poses a related question, which he promptly answers: "Why not have parachutists come here? The reason is *not* that it's about parachuting—or invasive species parachuting in. The actual reason is a market shift, but it's not an argument that's very well understood here. Once you begin down this slippery slope, what happens is that you start killing your past markets, your historical markets. Then we get back into the boom and bust [cycle]. You've busted your market. Once you lose your market, you will destroy tourism here." This is exactly what led to the collapse of the lobster and sea cucumber industry. One indicator of change is the type of tourist who now visits the islands. Most are older, wealthier, and more interested in being entertained by nature than in exploring its curious details. Call it "ecotourism lite."

Today, about five flights a day come to the islands from the mainland. More cargo ships are now needed because there is a food shortage. The number of tour boats has also grown exponentially. "Seventy percent of the fuel is actually used by the tourism industry," Watkins says, "and it's subsidized." In the long run, he believes that meaningful change must include dialogue among local residents who are still resentful at being shut out of the market. I tell Watkins that I think they have a point, and he replies: "Exactly. You need to understand all of these interests and be able to communicate with them. Everybody is right. That's the problem. That's the challenge."

Felipe Cruz is smoking a cigarette at a picnic table just up the road from Watkins's office. A light breeze blows in from the bay, carrying the smoke up in spirals. Cruz is the director of technical assistance at the CDF. He's also a brother of Eliécer Cruz, who was ousted from his position as director of the Galápagos National Park under the Gutiérrez administration. Felipe, who sits opposite me, stares out at the bay, taking another drag off his cigarette. "See that sailboat with the tall mast?" he asks. "I love that boat." He's referring to a sleek gray sailboat, streamlined and unadorned, and appropriately named *The Ghost*. Felipe is a tall, fit man with wavy black hair and a childlike grin. He's dressed in faded denim shorts and a white polo shirt with the word "Staff" on one of the sleeves. He owns his own sailboat in Puerto Ayora. Once he delighted his eight-year-old son by sailing back and forth across the bay while a playful sea lion chased a rope attached to the stern.

Felipe Cruz is a Galapagueño who was born on Floreana in 1958, when only about 40 people lived on the island. He is one of 12 siblings, all delivered by his father. The patriarch of this large family was able to do many things: He grew fruit trees and raised livestock with sweet water from a nearby spring. He fished the once plentiful waters around Floreana in his handmade boat. He was the town authority who later donated his home as the island's schoolhouse. Back then there were no roads, no televisions, no telephones, no neighbors. "It was a subsistence type of life," says Cruz, stubbing out his cigarette in a glass jar. As a boy he rode his horse around the thorn-scrub landscape, hung out with his siblings and friends, and was so obsessed with bugs that he was nicknamed the "boy of the insects." When a scientist came to visit his father, Felipe turned his interest to birds. He helped catch and study finches then released them back to the wild. When he left Floreana at the age of 12 to attend high school in Quito, he'd never seen a car. He adapted easily to city life, but the budding ornithologist was grounded in nature, and conserving it would become his lifelong passion.

After high school, Cruz became a naturalist guide for a few years, but then left the Galápagos to attend college in the United States. He received a Fulbright fellowship in the 1980s and returned to the islands to work as a field assistant at the CDRS. His first project was to help monitor the dark-rumped petrel and save it from extinction, which succeeded beyond anyone's expectations. Cruz then worked with the Galápagos National Park as head of protection and later as deputy director. In 1997 he returned to the research station to draft the Special Law for Galápagos, an ambitious proposal designed to give residents greater autonomy on the islands. Part of this Herculean task involved Cruz's role in conflict resolution within a divisive and often angry community. A major component of the law that was passed in 1998, and one that did not sit well with fishermen, was a paragraph extending the protected marine reserve to 40 nautical miles. Later, Cruz became technical director of Project Isabela, the largest island restoration project in the world, which succeeded in eradicating 100,000 feral goats in prime tortoise territory. But we'll come to that later.

I ask Cruz if he thinks the Special Law for Galápagos has been effective, despite its attempts to provide locals greater autonomy in the province. "Unfortunately, it was way ahead of its time," he replies, adding that he "masterminded" the project. "We, the people of Galápagos, were not ready to accept all the responsibilities the law gave us, which is a pity. The law remains one of the best pieces of legislation for any given protected area in the world, and Ecuador should get proper credit for passing it because we had to change the Constitution to get it. Not a single other country in the world has done that."

Cruz maintains that the visitor sites in the Galápagos are still the best managed anywhere, and that the park has done a good job in reducing impact despite the quantum leap in the number of visitors. The problem, he says, is that the tourist boom hasn't been regulated and the towns cannot expand to accommodate it. Even so, Cruz remains optimistic. "We have been recommending that the government of Ecuador needs to have a strong leadership in the Galápagos [and] this is the first government in the history of Ecuador that recognizes that Galápagos has problems." All former administrations have tried to hide the problems, he adds, "trying to cover the sun with one finger." Like Watkins, he believes that the flood of colonists dilutes any sense of community and that most newcomers lack both knowledge of and concern for the very resources that make the islands special.

"Let's be totally honest about it. Most of the Galapagueños are totally dumb in that regard. They don't see the problems. They just see greed and want to have more access to business. There are very few Galapagueños that have a proper view of what the hell is happening here or that care about it." He goes on the explain that what's needed is a shared vision that starts in the towns and spreads outward. But environmental education has so far been a complete failure. "We should be teaching about Darwin and establishing an educational system that is more according to the reality of Galápagos. It seems to me quite disgusting that most of the kids in Galápagos are growing up in cement towns and they are attending [school] in a classroom that hardly has a window, and they're playing in a field that is cemented, and they don't know the islands, not even their own island. Obviously, we have the best educational resource you could ever think of. How many other places in the world can you see iguanas digging nests to lay eggs? In my ignorant opinion, those are the best teaching tools you could ever dream of, and if we don't use that, what the heck are we doing with education in Galápagos?"

In November 2008 Cruz received the U.S. Conservation Action Prize from the Whitney R. Harris World Ecology Center at the University of Missouri, whose scientists have long been active in the islands. (I would travel with one of them to Española to observe the waved albatross and other rare bird species during mating season.) The conservation prize recognizes "unsung heroes" working behind the scenes on conservation and restoration issues.

It's past noon when I bid Cruz farewell and leave the research station to walk home. I'm deep in thought when suddenly I hear laughter. Looking up, I see a long line of schoolchildren wearing knapsacks as their teachers guide them up the road to observe Lonesome George and the other giant tortoises that have made the islands so famous. They've come out of the classroom and into the natural world, but not one of them stops to observe the huge marine iguanas basking in the sun on this well-traveled road.

CHAPTER 6

Tortugas

"It is not how much you do, but how much love you put in the doing."
—Mother Teresa

Quinta Playa is a strip of white sand crowned at each end by dark lava bluffs. The sea is turquoise green, a conveyer belt of waves that carry with them one of the most sensitive Galápagos species: the Pacific green sea turtle. The beach is sculpted with bowl-shaped depressions—the nests of turtles that came ashore to lay their eggs in a ritual that has continued for millions of years. It's a natural museum, a place where pink coral has fused with black basalt, and long red mangrove seeds appear in the sand. Then, too, there's the sleek black wing feather of a frigate bird; the pad of a prickly pear cactus; and purple seashells polished smooth by the waves. American oystercatchers, with beaks the color of pomegranates, waddle down to the shore, pecking for crabs.

Patricia Zárate passes all of this almost without notice. She's racing the clock before sundown, surveying the beach for something much less idyllic: signs of dead or stranded sea turtles. Soon it becomes clear that Zárate, a marine biologist with the Charles Darwin Research Station, is not at all happy. In less than an hour, she and her teenage daughter, Mariantú Robles, find five dead turtles—one of them decapitated and rotting in the equatorial heat.

This beach on the southern shore of Isabela Island is one of the archipelago's most prolific turtle nesting sites. It's where Zárate has worked since 2000 to monitor the protected reptile—the only sea turtle to breed in the archipelago. Like all species in the Galápagos, the green sea turtle is vulnerable to natural threats like beach erosion and seabirds that can snatch a hatchling quicker than a sneeze. But Zárate and others have noticed an increase in unnatural threats as migrating turtles drown in fishermen's nets, are lacerated by boat rudders, or are decapitated by a growing number of tourist boats in the protected marine reserve.

Zárate is a striking woman from southern Chile with long black hair and a muscular torso. She's as tenacious as the species she studies. The work demands that she deal with sleepless nights, long hot days, pounding rain, and bloodthirsty mosquitoes. The biologist is used to this fieldwork, but she can't do it alone. She depends on unpaid volunteers who come from all over the world to help. They're the backbone of this project and the reason it works. Since 2000, when her work in the Galápagos began, she has worked with more than 500 volunteers, about 70 percent of them from Ecuador. In February 2008 I join a group of nine young men and women in what, for many of them, will become the most challenging experience of their lives: building a camp on a remote beach with wood scraps, tarps, tents, and a makeshift kitchen that will serve as "home" for two months.

It's 3:30 a.m. at the Puerto Ayora dock, and bats flicker on the humid air, chasing mosquitoes. I sit on a damp bench with a 19-year-old student named Pablo Mejía and his mother, Mariana. We're waiting for Zárate and the other volunteers to arrive with a truck full of camping supplies. Pablo, like most kids his age in the Galápagos, is still in high school. He's a second-time volunteer in the sea turtle camp, and he's proved he can endure some of the toughest field conditions imaginable. He's a big kid who doesn't say much. He's wearing shorts, a white T-shirt, sneakers, and a battered baseball cap turned backward. His mother's a teacher. She tells me that of the 5,000 or so kids in this town only two know anything about ecology; Pablo is one of them. Now, as he silently fidgets, I ask him what his goals are. "To attend college and specialize in marine biology," he mumbles. That's exactly what his mentor Zárate has done, and as we sit in the predawn darkness I point out a night heron stalking the dock. "They eat *tortuguitas* (hatchlings)," he says, removing his cap and using it to swat a mosquito.

We're off to a late start this morning because two of the volunteers have overslept. This isn't good: Time and the tides are critical for our drop-off point, where we'll carry a ton of gear over dangerous rocks to a campsite about a mile away. At 4:30 a.m. Zárate, the rest of the team, and the supply trucks arrive. The trucks are loaded with boxes of food, water dispensers, tents, mattresses, tarps, field equipment, and countless other things, all of which had to be quarantined for three days in a freezer at the Charles Darwin Research Station. Why? To kill any invasive species that might have sneaked into a tent or a pair of socks. The gear is loaded into one speedboat. Zárate, the volunteers, and I travel in another.

Our boat heads out past tour ships in the bay, the shimmer of their lights reflecting on the water like iridescent snakes. Beyond town the waves roll in like mercury. Most of the volunteers have been up since 2 a.m., sorting and loading gear. Now, a few of them try to sleep as the boat slaps the waves so hard we're launched from our seats. This will be one of the hardest days of the journey, a series of hikes back and forth to camp, transporting a ton of gear over basalt as sharp as daggers. One stumble on an algae-slick rock can result in a sprained ankle, a gashed head, or worse. But for now, these brave young volunteers try to rest.

Across from me, Zárate's on another wavelength as she rocks out with headphones to something on her MP3 player. We pass Santiago Island and several islets without names. Soon after dawn, a pod of common dolphins appears on the port side of the boat, leaping and surfing in our wake all the way to their feeding ground.

About three hours later Isabela Island comes into view, edged with white sandy beaches. Rain clouds hang over Volcán Sierra Negra to the north like ink-stained gauze. As we approach the shore in a rubber dinghy, a blue-footed booby dives beak first into the surf and emerges with a fish. Then a sea turtle pokes its head out of the shallows next to the boat, as if in greeting. It seems both auspicious—and dangerously close.

We land on a tiny pocket of sand, where we unload gear from the supply boat and begin the arduous task of hauling it down to the campsite a full mile away. But the tide is coming in fast, much too fast, and we must hurry. Zárate's daughter, Mariantú, a curvaceous 19-year-old with long, dark dreadlocks, floats a cooler full of the only fresh produce the park service allows here: potatoes, carrots, plantains, beets, and onions. Her younger brother, Cristóbal, hurries through the surf with a stack of plastic patio chairs balanced on his head while the waves snatch a T-shirt he's tucked into his pocket. A 17-year-old named María Fernanda from Puerto Ayora slips on the algae-slick rocks and gashes her shin. I scrape my arm and clean the wound with salt water. It's the only wash water these volunteers will have for the next two months.

In the afternoon we begin setting up camp in a clearing at a distance from the nesting sites. I join Mariantú and a volunteer named Fabián Puebla down the beach and try to move plastic drums of drinking water to camp. The supply boat captain has miscalculated and dumped this daunting load a mile past camp in the other direction. By now we're dead tired, parched, and as sunburned as the beets in our cooler. The rectangular water containers are just too heavy, yet still we try to edge them an inch at a time through the sand. If they were round, we could roll them. Finally, we give up and head back to camp.

Along the way Mariantú and I find two sun-bleached turtle skulls. We also see *a lot* of things that do not belong on a protected beach. A huge truck tire lies wedged in sand about two feet from a recently dug nest. A few yards away there's a motor oil bottle, a Styrofoam take-out box, a plywood plank imprinted with Chinese characters, a plastic hot sauce bottle, blue and red plastic clothespins, a rubber sandal, a green fishing line, a coil of nylon rope, and part of an illegal purse seine net.

And then there are the bleach bottles, at least ten by my count. It's well known that some Galápagos fishermen squirt the poison on squid, which stirs them out of their hiding places. Squid emit biolumines-cence at night, attracting hungry sharks and sea lions. These latter two species are lucrative catches for the black market in Asia: shark fins are prized for shark fin soup, and sea lion penises are considered an aph-rodisiac. These thoughts weigh heavily on me back at camp, where we eat a modest dinner and try to rest before darkness descends and our work begins.

———————————

Except for a scratch of moonlight in the southeastern sky, it's pitch black on the beach. Waves tumble in, carrying with them dozens of turtles. The beach is a mosaic of tracks, some crossing over others or cir-cling back toward the sea. Zárate traverses the beach with a headlamp covered with red cellophane, looking for nests. Sea turtles are easily disoriented by artificial light. Light in the red spectrum is invisible to them. She stoops to examine a recently dug hole with an empty egg chamber. "This is interesting," she says. "She abandoned the nest with-out laying her eggs. Turtles do that if they don't find a suitable site," she explains. They can return to shore up to eight times over a two-week period until conditions are right.

As we continue along the beach our footprints release bioluminescence in the sand—the light of living organisms. I feel as though I'm walking across the galaxy. Soon Zárate spots a female turtle returning to the sea. The biologist kneels in the sand and straddles the turtle's head between her thighs. *"Tranquila, mi hija,"* she tells the struggling reptile. "Calm down, my daughter." Fabián, the volunteer who tried to help move the water canisters down the beach, kneels in the damp sand, pulls out a pocketknife, and removes huge barnacles attached to the turtle's back. Zárate wants to take them back to the lab and analyze where they came from and whether they contain parasites.

A slender young volunteer from Brazil, Manuela Borja, measures the turtle's width and girth. If a turtle hasn't been tagged, Manú, as she's called, punches a hole in one of its flippers and attaches an identification tag. This makes it possible to keep track of returning females—and new arrivals. It also indicates how long they've been out to sea and reveals the fate of turtles caught in nets or deliberately killed. In 2005, for example, Zárate learned that the turtle with tag number DC235, one she had tagged, had been caught in the longline of a commercial tuna boat off the coast of Panama. The fishermen could have set the turtle free. Instead, says Zárate, one man kept the tag as a souvenir after they killed her and ate her for supper.

Under international law, commercial fishing is prohibited in the Galápagos. But green sea turtles, like other species, migrate long distances to feed. To get to these foraging grounds they become vulnerable, especially in the coastal waters of other Central and South American nations. Research conducted by Zárate and others between 2002 and 2007 shows that the main causes of death among the Pacific green sea turtle are shell fracture, heat stress on nesting beaches, being stranded after injury from a boat strike, an embedded fishing hook, and intentional amputation of head or limb after a turtle becomes trapped in a fisherman's net. The majority of turtle deaths—64 percent—are caused by fishing and boat gear. Many die naturally or from disease. The causes of about 27 percent of sea turtle deaths are unknown.

Zárate and longtime colleagues like Jeffrey A. Seminoff, a marine ecologist and leader of the Marine Turtle Ecology and Assessment Program at the Southwest Fisheries Science Center in La Jolla, California, and Peter H. Dutton, a marine biologist with NOAA Fisheries and the head of the Marine Turtle Research Program, believe that the species can survive only if international laws on sea turtle conservation are strengthened and enforced. But this will require a cohesive and long-term vision among biologists, wildlife managers, economists, policymakers—and, of course, fishermen.

———————————

Back on the beach it's almost midnight when we come upon a pair of intersecting tracks. "OK," says Zárate. "One has come up the beach, and the other has returned to the ocean." She sends Fabián up the trail to look for the new arrival. From where I stand in darkness, he's a silhouette against the night sky. I follow. The turtle has found a suitable site and is using her flippers to dig a nest and a narrow egg chamber. Fabián and I lie on our bellies, peering down into the nest that is being dug. Sand flies into our faces like glitter. In the heavy gravity of terra firma, it's clear the turtle is tiring as she throws her head back and gulps for air. When she completes her nest, we shine a red-tinted light on her vulva. Soon, a slimy liquid emerges and the first egg plops into the foot-deep chamber. It takes about 30 minutes before the final egg, number 44, falls. Now near exhaustion, she uses her flippers to cover the eggs with sand, turns, and plods back to the buoyancy of the sea.

In the distance to the east, the lights of the fishing village of Villamil, population 2,500, appear strangely out of place. They are far enough away that they won't disorient sea turtles; Quinta Playa, the fifth beach west of Villamil, is very dark. At other nesting beaches the light from cruise ships has caused arriving females to turn around and return to the sea. Zárate has documented this over the years. "I don't know why they have to keep those lights on," she complains. "People [on board] are dreaming, sleeping. They don't need lights."

Curiously, sea turtles don't have sex hormones. The temperature inside the nest determines the sex of the hatchlings. This phenomenon fascinates Manú, the bright young Brazilian volunteer I worked with alongside Fabián and Zárate. In fact, Manú first became interested in sea turtles as a child and plans to work on an advanced degree in sea turtle conservation. She has volunteered in turtle camps before, and one afternoon after a dip in the ocean, she approaches me at my tent with some research papers she has brought from Brazil on the sex ratio of sea turtles. The papers indicate that if the eggs incubate below 86°F, more males tend to be born. At temperatures higher than 86°F, a nest can yield all females—a potential reproductive nightmare—if global warming continues. I've snorkeled with sea turtles all over the Galápagos; the graceful strokes of their front flippers make it seem as if they're flying through water. But it's the eye of the turtle, black and round and alien, that has humbled me most and has made me feel like an aquatic interloper traversing an endangered world.

To many traditional cultures the turtle is so sacred it's considered the Creator of the World. Several indigenous groups in North America relate a story about the Great Turtle. Long ago, before humans inhabited Earth—before Earth was even born—Sky Man uprooted a giant celestial tree. His wife peered down through the hole it made and saw stars glimmering in darkness. Then, bending closer to examine this strange new world, she fell through the hole into starlight. The continents did not yet exist, only the sea. Winged creatures from the ocean looked up and saw her falling. They called her Star Woman and flew up to catch her. Then they set her on the back of Great Turtle, who was swimming in the ocean. Because there was no land, the Water Creatures dove down and tried, one by one, to bring mud up and cover Turtle's back. Only Muskrat succeeded. While Star Woman slept, the soil on Turtle's back changed. Seeds sprouted. Trees flourished. Flowers appeared in a kind of paradise: Turtle Island, the great Mother Earth we all ride on. Star Woman then gave birth to a female child named First Woman, whom West Wind impregnated. She gave birth to twin sons who named all the living things on Earth.

———————————

It's the third night on Quinta Playa and Zárate can no longer walk the beach. She has pushed herself too hard, and a blister the size of a dime has formed on one of her heels. Reluctantly, she decides to stay in camp while the volunteers depart in teams for the next eight hours of observation. I join her at a makeshift table surrounded by the plastic chairs her son had carried on his head through the rising tide. Zárate winces as she rests her injured foot on one of them. A single candle, stuck to the top of a can, provides the only light. I ask her about one of the greatest mysteries surrounding sea turtles: What compels them to return to the same nesting beach where they were born?

"The magnetic compass of the Earth," she responds. "Some people also believe they can smell the composition of the island, they can smell the different minerals" in the sand.

In his *Voyage of the Turtle*, Carl Safina, an award-winning naturalist, confirms that sea turtles have an extraordinary sense of smell, and that they can respond to changes in Earth's magnetic field to create an inner "map" of where they are going. Satellite images have proved this. "What I find stunning," Safina writes, "is that the little hatchlings spring from their eggs as fully functional global-orientation systems, all booted up and running within their first minutes out of the sand. When they leave the beach for what, to us, is a great unknown, they don't even need to know what to do; it comes with their autopilot."

Safina is one of Zárate's heroes, but one of her earliest role models was the zoologist-host of a television series called *Daktari,* which aired in the late 1960s. The fictional doctor, Marsh Tracy, is a veterinarian who runs an animal research center in an East African wildlife preserve called Wameru. But the star characters in the series, whose storyline centers on protecting wildlife from poachers, are the doctor's pets, Clarence, a cross-eyed lion, and a naughty chimpanzee named Judy. "I used to say, 'I wanna be like him. I wanna be a veterinarian,' " Zárate recalls.

Throughout her childhood she and her family often spent weekends camping in the Chilean countryside. When they stayed in their cabin, she indulged her curiosity about the natural world. "I remember collecting frogs and insects and putting them into jars. Every time I saw an animal at the market I took it back home. I had parrots and a goat and a fish and little chickens. I had cats and rabbits and, of course, dogs. Those were my favorites." Zárate got her first dog when she was two years old. The family named it Laika after the dog sent into space on Soviet Sputnik 2. "She was the best friend of my life."

As a college student in Chile, Zárate learned how to free dive—without scuba gear. Her scholarship required her to participate in a competitive sport called *cacería marina* (underwater hunting). You hold your breath, go down as deep as you can with a harpoon, and kill fish. "To eat?" I ask. "No, just for fun. I didn't want to kill animals, but I had a scholarship and they gave me breakfast, lunch, and dinner, and some money to keep doing it." But the ocean and its mysteries had captivated her. "It's like being in space, free. That, besides the fish, the colors, the animals, the way they move, the way they look at you." She decided to become a marine biologist, and in 2000, the U.S. National Marine

Fisheries Service (a division of the National Oceanic and Atmospheric Administration, based in La Jolla, California) offered her a job as a volunteer in the Galápagos to monitor Pacific green sea turtles. Lack of funding at the Charles Darwin Research Station had left a 17-year gap in research on the species.

Zárate's first task was to bridge that gap and establish nesting, population, and migration patterns for the years 1983 to 2000. She and others found that the greatest threats to Pacific green sea turtles in the Galápagos were natural: beach erosion from high tides; warmer ocean currents during an El Niño year; or predators, such as frigate birds, lava gulls, herons, boobies, pelicans, beetles, and crabs that prey on the hatchlings as they run the gauntlet to the sea. But the obstacles didn't stop there; hungry sharks and other fish awaited them out at sea.

Zárate's blister is bothering her and she shifts in her plastic chair, her face illuminated by candlelight. The biggest shock to her *now*, is the growing number of *unnatural* threats that have proliferated during the last couple of years; leaving many turtles without heads or limbs. "Someone mutilated them maybe because they got entangled in a net, and for a fisherman, the net is the most valuable thing." Studies also reveal a rise in the number of turtles with fractured shells and atrophied or missing limbs. The number of tour boats in the reserve has increased dramatically, and they often collide with migrating turtles—as our own speedboat nearly did as we arrived on the island. But the species is a survivor, says Zárate. "The instinct is so strong that they will crawl up the beach without legs, or with bleeding wounds. I have seen nesting females with only one leg or one so atrophied she can't really move it." It takes these turtles twice the time—or longer—to build an egg chamber. If it collapses, they'll move to a different site and try again until their eggs are safely deposited in the sand, where they incubate for about 60 days before the hatchlings emerge.

"As a scientist, do you have to keep an emotional distance?" I ask.

"Yes, feeling sorry will take you away from finding a solution."

One solution, she says, is teaching Galapagueños why it's important to protect the sea turtle and its environment. That's why she works with high school and college students. The next step, she adds, is a series of workshops with fishermen.

Zárate retires to her tent for some well-deserved sleep. That night it rains. Hard. We've pitched our tents under a communal tarp called a *techo,* but the rain blows in from all directions. It's hard to sleep, and as I sit up in my tent I notice a dark figure walking around camp. She's wearing a red headlamp and an ankle-length raincoat. It's one of the volunteers, poking the now sagging tarp with a stick to drain off the rain so it doesn't collapse on us. Soon she appears right outside my tent. As her pole nudges the tarp, buckets of water spill directly onto my head. I'm too soaked to sleep, so I lie awake for hours listening to birdsong from a salt marsh just beyond camp. Eventually, it sends me off to sleep.

By sunrise the rain has finally stopped. I wake to find Pablo digging holes in the sand to create a recycling center for cans, plastic, and paper. Organic scraps are dumped into a deep hole on the other side of camp, attracting scores of hermit crabs. Pablo is a true workhorse. He hasn't slept a wink, and when the recycling posts are sunk, strung with plastic bags, and labeled, he sits at the camp table with Zárate recording last night's data. Most volunteers are still asleep; Fabián is snoring in the tent next to mine.

After breakfast Zárate reminds the volunteers what their duties are. "Everyone needs to respect each other, to do their share," she tells them, like sorting the trash, cooking, washing dishes, recording data, and checking the techo for water. "Here it can rain even harder than it has. I've seen the rain take down the tarp over an entire camp, the techo was so heavy with water." The volunteers hang their heads. "But thank God you've all been so strong. For this I respect you." They're dead tired and not smiling as they scoot out of their chairs to begin another day's work.

I wander off for a while to explore the lagoon, home to a variety of birds, including the greater flamingo. At one time, I'm told, more than 300 flamingos lived in this marsh. Now, only about 30 exist here. Against the dark background of brackish water they're psychedelic pink. The graceful movements of their legs, their curved necks, their black-tipped beaks make them look as though they're writing calligraphy in the mud. Flamingos are pink because of what they eat: brine shrimp and other rosy-colored crustaceans. Suddenly, a flamingo takes flight, its head extended forward, straight as an arrow. It looks ungainly, yet there's an elegance in its movements as it soars out over the mangroves.

By late afternoon the sky is an enormous bruise, promising more rain. While some of the volunteers bathe in the ocean, I walk down the beach to a rocky outcropping past a growth of beach morning glory, inkberry, and Galápagos purslane. Just offshore I see about ten dark disks floating on the water: turtles. Every once in a while, one pokes its head out of the water then disappears. They're waiting for darkness to descend before coming ashore. Just past 5 p.m., I notice a young female crawling up the beach in broad daylight. She has chosen an abandoned nest next to the truck tire, but she soon leaves it to dig another. Fabián and a volunteer named Jakob Turtur, from Munich, run to camp and grab their gear. The turtle digs and digs. Sand covers her head and clings to the mucous that drips from her eyes like tears. The eggs keep falling, one after another. The count when she's finished: 85 eggs. We've just come face to face with one of the oldest species on Earth and witnessed in broad daylight its astonishing will to survive. Charles Darwin could not have explained this phenomenon more clearly than when he wrote in his journals in 1845, "one is astonished at the amount of creative force ... displayed on these small, barren and rocky islands."

Back in camp that evening Zárate cooks dinner on a two-burner stove. The volunteers sit at the table playing a card game called *Idiota* (Idiot), also known as *Cuchara* (Spoon). Here's how it works. The dealer gives everyone three cards. Then the person to your left hands you a card as quickly as possible. When someone gets three of the same number, he or she grabs a spoon from a pile in the center of the table. The other players must do the same. The last one to grab a spoon gets the first letter of the word "idiota" written on his or her forehead. The game continues until the forehead of one of the players bears the entire word. This happens to María Fernanda; the camp overflows with laughter. It's the first time I've seen these young volunteers so happy.

That night, well fed, they gear up and head down toward the sea, their voices trailing off until only the surf prevails. Tonight they'll take blood samples from turtles with scientists from the Galápagos National Park to test for parasites and other ailments, drawing the dark liquid into tubes for analysis.

Mariantú joins workers from the Galápagos National Park at a different beach. It's the first time such tests have ever been done and analyzed. In total this season, she and others will take blood from 96 turtles on Quinta Playa, and more than 100 on two other nesting beaches, Bahia Barahona on Isabela Island and Las Bachas on Santa Cruz Island. But drawing blood while a female is laying her eggs can prove disastrous, she tells me.

"Even if she has laid one egg, if she feels the needle she'll stop and abort. She'll begin covering the egg chamber and return to the sea." The group began taking blood samples instead just as the turtles emerged from the ocean or just before they returned to it. "Giving birth," Mariantú says, "is just too special. It's like, 'Congratulations. You're going to be a mommy.' "

After five days in camp it's time for me to leave. I have work to do elsewhere, but I'm sad that I won't be there when the hatchlings poke through their shells and scramble to the sea. When my speedboat heads into the waves I can see Zárate and her volunteers waving good-bye. They're out on the beach picking up trash, creating passage for all those mommies.

In May I visit the sea turtle lab at the Charles Darwin Research Station, where a large picture window overlooks the bay. Metal shelves on one wall are lined with specimens preserved in formaldehyde, including a huge barnacle that will be tested for parasites and a hatchling that didn't make it. Zárate is working at her computer while a few other volunteers help sort through the data. I'm surprised to see Pablo there with María, the girl who had ended up with the word "idiota" written on her forehead after the card game. María shows that she is anything *but* as she works alongside Pablo, transferring the data they collected at Quinta Playa from a logbook into a computer.

"*Hola!*" says Pablo, surprised to see me. The results of their studies are still coming in, and I want to hear about them. Zárate and I go outside and sit on a stone wall. "I think this is going to be the best year since I've been here," she tells me. The biggest surprise, she adds, is that 15 percent of the turtles were "re-immigrants," tagged turtles that had returned. The other 85 percent were new arrivals. In all, the volunteers tagged more than 9,000 turtles this season on three different beaches.

Preliminary tests from the blood samples revealed the presence of parasites from mosquitoes. But they also showed high levels of defensive white blood cells the immune system musters to fight infection. The data are incomplete, though, and Zárate must get back to work. She was just accepted into the Ph.D. program at the Archie Carr Sea Turtle Center at the University of Florida in Gainesville. Her research there will focus on the ecological role of sea turtles in the Galápagos.

"So you'll be coming back?"

"Yes, my plan is to continue working with sea turtles in the Galápagos," she declares, as certain as the rising tides in the bay just behind us.

———————————

And what about those other brave volunteers who made her research possible? I want to know more about their stay at Quinta Playa, so I contact Manú and Fabián, the two I accompanied down the beach during the first week of monitoring. Here's what they recall:

Fabián: *The first days were very hard because our bodies had to adapt to the environment just like the other species that live there. The stifling heat in the morning and afternoon, the strong winds, cold nights, the mosquitoes and small biting flies characterize this time of year. Even so, my desire to work with this exceptional species was worth it every single day in camp. Some nights we worked between eight and ten hours only by the light of the moon, or with red-tinted lanterns. Sadly, some of the turtles arrived on the beach after suffering tragedies while migrating through the Pacific Ocean to their nesting grounds—scars on their shells or bodies from boat rudders or encounters with sharks. Even so, it will be impossible to forget the satisfaction we felt the first time we watched a female deposit 70 eggs. There won't be a single day in my life when I don't think of what we achieved on Quinta Playa. I hope that other countries also work to protect this fragile species that's in danger of extinction.*

Manú: *After Paty [Zárate] left the day after you did, she asked me to be responsible for the camp, to be the leader, you know. This was not easy because at other turtle camps I'd only been a volunteer. The most difficult thing for me was dealing with the relationship between the volunteers and being responsible for [delegating] the work. Everybody was very different from each other. Sometimes I didn't know how to speak with them, how to ask, how to explain, and I just wanted to go home. But in the end, we got along very well. I was impressed with the group's determination and the amount of data we collected. After a while I felt at one, as though every plant, every animal had became part of me. When the time came to leave my heart started to cry because I felt I would never be part of this beautiful family again: The turtles, the birds, the sea lions, the flamingos, the rocks, the sea. But they're with me all the time. They help me live, breathe, and fight for their survival.*

A few weeks later in Puerto Ayora, I meet up with Zárate's daughter, Mariantú. She's carrying the seedling of a muyuyo tree, known for its bright yellow flowers. She wants to plant it before she and her mother move to Florida. Her dreadlocks have turned two shades lighter, and the diamond stud in her nostril seems brighter, too, against her sun-bronzed skin. I invite her to my house to catch up.

So a few days later she stops by with her black dog, Mara, a sweet little mutt with tufts of white on her legs. Now that Mariantú is leaving the islands, she's scouting out a good new home for Mara. "It's so sad," she says, her brown eyes brimming. "It will be like a knife in my heart." I give Mara a bowl of water and Mariantú a glass of pear juice. I also give her a hug: I know what it's like to lose a dog. We sit on my couch and chat.

Mariantú has been working in turtle camps for at least six years. She says her mom has come to rely on her because of her experience in the field, especially in the Galápagos, where she moves from one nesting beach to another. "Here it's really different from other places. First, the camp is in the wild, with nothing else. In other [countries], you have a bathroom, a kitchen, a person who cooks, you have lights, or you're near a town. But here you have nothing. The conditions are extreme.

"For two months, there are no fresh vegetables or fruit, no cold water to drink or to take a bath." There's also a risk of accidents, say, a broken leg or a head injury, or an infection that needs immediate treatment. Communications are sketchy because the camp is so remote. The primary form of communication is by cell phone, or VHS radio, but the latter is even less reliable.

I ask if the camp conditions troubled the volunteers.

"Yes, obviously," she responds. "But the important thing is that in spite of all that, you're living on a quiet beach, a wonderful beach, a beautiful beach. You see the animals all day long. There are no worries. It's like being in a bubble, out of the world, and you want to stay there. You get used to the mosquitoes, the rain, living in a tent, lack of sleep, but when you come back to town, you spend like one day; all you want is a glass of cold water, and then you want to go back to the camp."

Toward the end of their stay on Quinta Playa, she says, up to a hundred turtles were coming ashore every single night. Her long black eyelashes blink in amazement. "*Chuta!*" she says. "Wow!" "There were highways on that beach!"

When volunteers had free time they read books, juggled balls, practiced yoga on the beach, played cards, bodysurfed, and learned to cook. Some even made chocolate cake with a surplus of flour and cocoa. "You become really, really good friends." Mariantú stares out the window and says how strange it is being back in town.

"You can't drink water out of the tap [here]. You don't have medical service. If you are really sick, you need to go to the mainland. You cannot get hurt there. You cannot have an accident." Then, back in town, all the traffic, the noise, the constant influx of tourists is almost too much to bear. But her biggest lament is the lack of education in Puerto Ayora, especially a university where she can expand her knowledge. Even the high school, she says, is completely inadequate. "That's why my brother left [and returned to Chile]. He wasn't happy with the educational system here; he wanted more options." So does Mariantú. "I have aspirations," she declares. "I want more opportunities. I want to meet people with more experience."

This year in Gainesville, while her mother begins her Ph.D. program, Mariantú will attend a community college to begin studying animal science. Then she'll pursue her lifelong dream of becoming a veterinarian, just as her mother has persued hers. But now a painful task awaits her as she attaches a bungee cord to Mara's collar and walks her out my gate to meet with a prospective adoptive master. As I watch this budding vet pass beneath my acacia tree, I know she'll be absolutely perfect.

CHAPTER 7

Into the Blue

*"How inappropriate to call this planet Earth
when it is quite clearly Ocean."*

—Arthur C. Clarke

Few people know more about marine iguanas than Jack Nelson, former owner of the famous Hotel Galápagos. The hotel and main lodge overlook Academy Bay on a rocky shoreline that's still as natural as it was when Nelson's father Forrest sailed here in 1960 and built this hotel, the first in the islands. The grounds are a natural wildlife refuge where pelicans perch on lava etched smooth by the tides and where great blue herons stalk crabs. It's also a nesting site where marine iguanas have sparred, mated, dug nests, and laid their eggs in a stunning ritual that Nelson has witnessed for the last four decades.

Jack Nelson is something of a legend in Puerto Ayora, a brilliant thinker, a dreamer, and an expatriate who moved here from his childhood home in California to escape the Vietnam War when his draft number came up 1A. Today he's the U.S. Guayaquil consular agent for the Galápagos, a diplomatic role that requires intervention when Americans get into trouble or when their tour ships sink—and they do.

I first meet Nelson at the Hotel Galápagos in 2006 while I am on a teaching assignment on the islands. From the Hotel Mainao, where I am staying with my students, I walk past tourist shops, follow a walkway up to the lodge, and knock on a rusty screen door. Nelson, now in his early 60s, steps out to greet me: "Hello, hello. Please come in," he says. Nelson is a fit and handsome man with intelligent blue eyes and a gentle voice. He's dressed in a faded Hawaiian shirt, knee-length shorts, and hiking boots. He seems grateful to have a visitor; the hotel recently closed, but he still lives here. Sadly, the buildings of this landmark have surrendered to the salt, wind, and rain and are beyond repair. The options: tear down the structures and rebuild, or sell the place and let someone else figure it out.

Nelson leads me outside to his patio overlooking the bay. We sit at a weathered table shaded by a roof of palm fronds. Darwin's finches hop around in a cluster of mangroves, and marine iguanas bask in the sun just a few feet away. Nelson glances to his left: "Look at that marine iguana," he remarks. "There's a tick on its butt and he can't get to it." The reptile is a contortion of snout and feet, like a dog chasing its tail, but the reach is too far, and it soon gives up.

Nelson is the "go-to guy" in the Galápagos. There's a kind of transparency about him because he tells it straight. Like everyone here, he has his own opinions, but he's stayed the course despite the enormous upheaval created when the Special Law for Galápagos was passed in 1998. He even helped write it. He has been interviewed by just about everyone, from Japanese filmmakers to the BBC, to the *New York Times*, and most recently by television reporter Dan Rather, who came to the Hotel Galápagos with a small crew. Rather wanted to know whether the Galápagos can survive now that the islands are endangered. Nelson was diplomatic. He told the newsman pretty much what he tells everyone. Something like: *There's still time to make things work, but outside pressure and corruption among clever politicians, lack of education, and disparate values have always formed the Achilles' heel of what could have become a sustainable jewel instead of a myth.*

For most journalists, the main questions are about the survival of the Galápagos. For one thing, the Galápagos have no indigenous roots; most colonists are newcomers with little sense of identity, as Cruz pointed out at the Charles Darwin Foundation They've lived here only a few years. The question then naturally arises: What does it *truly* mean to be Galapagueño—to connect with the land, the wildlife, and the ocean—to love the very resources that sustain you instead of destroying them to get rich quick?

Herein lies the Galápagos koan—the question with no rational answer.

———————————

Jack Nelson was born in Lockport, New York, near Buffalo, in 1947. When he was four his father loaded his gear—and his family—into a small sailboat and tacked south to the Florida Keys and the Bahamas. Jack's sister, Christine, was seven. His mother, Bawn, a pretty young woman, served as just about everything as "first mate," and she prayed that her toddler, Jack, would not topple overboard. When the Nelsons returned to Florida, Forrest moved his family to California but soon sailed away with a buddy, down the Baja coast to Mexico, the Pacific shores of Central America, and all the way to the Galápagos. The year was 1952.

"He'd be gone for long periods. Sometimes we wouldn't see him for a year," Nelson says. "He'd go sailing or work on shrimp or tuna boats." Meanwhile, Bawn realized she'd married a charming wanderer with great maritime skills and an addiction to the sea. In her husband's absence she began working at secretarial jobs or at places like Douglas Aircraft in Long Beach to support the children. "It was your typical dysfunctional family," says Nelson. "He [Dad] didn't send much support so we moved around a lot." When Nelson graduated from high school in Long Beach in 1965 his life, like that of many young people, took a complicated turn.

"What changed?" I ask.

"In a word, psychedelics. After I graduated I had a couple of odd jobs, let my hair grow, wore weird clothes, hitchhiked around, got high a lot."

Nelson's first trip to the Galápagos was in March 1967. He returned to the United States in August 1967 by hitchhiking from Panama. Then he returned to the islands in September 1968.

Somewhere in between, though, lurked the Vietnam War and the draft. "I decided it was either Canada or Galápagos, and my father had already moved here in 1960. What I saw was that the Vietnam War was being pursued so stupidly. The way the U.S. was dealing with the Vietnam question and the Southeast Asia question was so brain-dead, I just couldn't participate. I thought, it's so destructive to American interests that I said 'No, I'm not gonna do it.' "

Eventually his number was called and, like his father, Forrest, Jack Nelson became an expatriate. The Galápagos provided the perfect lair for eluding the authorities. Nevertheless, the Draft Board knew where he was living and had sent him about six draft notices. The one that contained an indictment against him was sent to the Galápagos in 1970, but he never received it. "That was the year the mail boat sank," he explains. He sent an official sworn affidavit back to the United States. Case dismissed. After President Richard Nixon resigned, in the summer of 1974, President Gerald Ford granted amnesty to about a half million draft dodgers under a work compensation program. But the year was 1975 and the United States was in the throes of a job crunch. Even returning war veterans couldn't find jobs, let alone those who had fled.

So Jack Nelson stayed on, but like many adventurers, wanderlust tugged at his soul and he soon grew weary of the Galápagos. There was almost nothing in Puerto Ayora, the one-horse town he called home. For the next few years he traveled, doing odd jobs, working on boats and in the import-export business. He helped navigate a yacht to Acapulco and then to various islands, including a small coral atoll called Ahe in the South Pacific. After that, it was Tahiti and American Samoa, where he landed a job in a grocery store to make ends meet.

"You became your father's son," I remark.

"Well, footloose, that's for sure. I was just wandering to see what was out there."

But when Nelson had returned to the Galápagos in 1968, he learned that in his absence the year before, his sister, Christy, had married the hotel foreman, a charming visionary named José Luís Gallardo who owned a ranch up in the highlands and eventually began practicing artificial insemination among his livestock. "I was both surprised and delighted," Nelson says of his sister's marriage. In 1968 there was still no tourism other than the occasional adventurer and the charter clients of Fritz and Karl Angermeyer, who usually stayed on those two sailboats and traveled for weeks around the islands. In September 1968 Forrest traveled to Quito to talk with Hernán Correa and Eduardo Proaño of Metropolitan Touring to convince them of the possibility of running tours in the islands. Forrest already had a complete plan. Metropolitan would take on the receptive end of management and loan him money to expand the hotel to 12 rooms. The Hotel Galápagos would charter a boat and an airplane; Metropolitan Touring would handle the mainland hotels and transfers. The Hotel Galápagos would pay the loan out of profits.

At the time, there were no regular flights to or from the mainland, and passengers would often be put up in the military barracks on Baltra Island near the airstrip to wait. Eventually an old bomber fitted with jumpseats arrived. "My dad bribed the pilot 50 bucks to let him on the plane; somebody else gave the pilot a baby goat," Nelson recalls. The plane took off and ascended to 28,000 feet, but there were only four oxygen masks. "They were trying to give the goat oxygen—the navigator was sharing his mask with this goat—and the goat was just not making it," says Nelson, laughing out loud. "The goat's going bahh…. So they drop back down to like 12,000 feet and fly to Quito that way."

Forrest obtained the loan and immediately got to work building eight new rooms. The first tour group came to the new Hotel Galápagos in April 1969.

Jack helped his father run the hotel over the years and entertained such famous guests as Prince Philip, the Duke of Edinburgh. The prince arrived on the royal yacht *Britannia,* with a large contingent of princesses and Her Ladyships and other officials. Lord Mountbatten was also in the party. It was 1974 and Ecuador was under the military rule of President Rodriquez Lara, who arrived at the hotel with his own party. "All together we had about 85 people there for lunch," Nelson recalls. "At the time there was nothing here. You didn't just go to the market and get a bag of carrots, or lettuce, or anything else. It was very difficult to get supplies."

Nelson was in charge of cooking roast beef for all those visitors, but he had access to only three small ovens, including one on a boat anchored out in the bay. So he zipped back and forth in his speedboat to marinate, baste, and roast the beef. In the end, all turned out well. "It was quite a show; everybody was pleased," he recalls. "The British royal party left immediately after lunch, and the Ecuadorian military [the generals, mayors, and corporals] stayed on until late in the night, polishing off bottles of whiskey and vodka."

Jack Nelson kept trying to leave the Galápagos. He lived for awhile in Hawaii, where he learned the art of scrimshaw, and then Alaska, where he tried to make a living fishing. Ultimately he moved back to California. "One of the lessons I've learned is that after you've lived in the Galápagos for too long, you're not competent for anything else. You build a set of habits and skills and knowledge that are applicable here, and it's such a unique place that if you try and take those skills somewhere else, you may or may not make it." His life reminds me of the final lyrics from the Eagles song "Hotel California": *"You can check out anytime you like, But you can never leave!"*

The hotel itself is a testament to Nelson's creative spirit and the years he's spent at the historic Hotel Galápagos, which is like a museum. Near the entrance hangs an old map of the islands, yellowed and faded with age. Puerto Ayora has become a smudged fingerprint from all the guests who have pointed to their location on the map. The bar still stands, lined with empty bottles but not a single customer, like a scene out of *The Shining*. Nelson's oil paintings surround the room. One shows an enormous volcanic eruption in 1825 on Fernandina Island as a ship, the *Tartar*, passes below in ghostly red.

Mobiles hang from the ceiling, elegant works that Nelson made from fan coral, a fibrous and colorful species that resembles lace. Nelson found the coral while free diving back in the 1970s when no one considered the consequences of such actions. He cleaned them up and used filament fishing line to string them to a piece of driftwood. There's also the skeleton of a pelican that his niece Corina brought him one day on the end of a stick. "I buried it in a wet spot in the sand for about six months, and then we uncovered it and cleaned it up. I put it together with hot-melt glue and copper wire to get the bones to mount to simulate an actual flying position." It's a graceful masterpiece. When a professor of taxonomy came to visit she was duly impressed. "I like that," she told Nelson. "My graduate students put birds together so that they look like 747s. They look so rigid. But this one's got a live twist to it."

Feathers and fish hang from another mobile. Nelson made the fish from copper window screening that he scavenged off Baltra Island, where the U.S. military had an air base during World War II. The buildings were gone but big sheets of copper screening littered the ground. On a wall near the bar hangs a giant tortoise carapace from the highlands of Santa Cruz. It bears a bullet hole near the heart level. "Until the 1930s and '40s the tortoises here were still regarded as a food source," Nelson explains. The tortoises also provided a plethora of very rich fat that the local colonists collected and rendered down for cooking oil. "It was quite common to slaughter the tortoises back then."

"What are those bones?" I ask, pointing up at several large objects.

"The one on the left is the skull of a pilot whale. That's a whale that gets up to about something like 16 feet or so. A friend of mine gave me that when he found it many, many years ago on Punta Roca Fuerte, which is the easternmost point of Santa Cruz Island." Also on display are the vertebra, skull, jawbone, and rib of a sperm whale. But the rarest natural object in the hotel is the skull of a goosebeak whale. Nelson didn't know what it was until a cytologist came through who was an expert on the species. "He said, 'Absolutely, that's a goosebeak whale. How exciting!' "

Among the furnishings of the hotel are lamps made from driftwood and the skin of cactuses, which emit a soft orange glow. Natural objects like these cannot be taken from land in the Galápagos National Park; they must come from private property. That makes it difficult for Nelson and his sister, Christy, an accomplished artist whose eclectic sculptures are made from found objects and sell for hundreds of dollars at the upscale Galería Aymara.

Nelson and I go back outside to visit with the resident marine iguanas. He's watched them so long he can identify each one individually. I, too, am growing fond of them and want to learn more. It's mating season and the now pregnant females are digging nests to lay their eggs. It takes about three months (95 days) before the eggs hatch and baby iguanas poke through the soft shells. The conditions must be perfect: soft earth or sand. I watch a few of them crawl around, searching for the right spot to dig. One female is particularly aggressive and protects her territory, hissing at any potential interloper: Iguanas will dig up another female's nest and cast out the eggs. A small female finds a good spot and begins digging with her front legs, then pushing the sand out with her hind legs. It's hard work; there's not much loose ground around here. A few of the younger females have even tried to penetrate a tough grassy patch near the ramada and failed. Nelson felt sorry for them so he covered that grassy area with a large mat woven from palm fronds to discourage digging.

As we sit talking about Galápagos wildlife, a yellow warbler with a red-tinged head alights on the table, unfazed. Nelson is a former naturalist guide and a natural storyteller. As he gazes down at the warbler, I can see that a story has come to mind. Once, right here on his land, he watched a great blue heron nab a yellow warbler with its long saffron-colored beak. The little bird screamed so loud that "a hundred or so other yellow warblers" flocked around the heron to see what had become of their kin. It was one of the most startling natural events he'd ever witnessed in nature.

The iguanas have now settled on the rocks and are soaking up sunshine. I ask Nelson why he loves this species so much. "It's the only marine iguana in the world. For one thing, they're so approachable. You can sit down with a bunch of them and they're not upset." As for Charles Darwin's description of marine iguanas as "imps of darkness," he totally disagrees: "I never see them in that cartoony sense. I see a real grace in them. They're perfect for what they are."

In 1995 Nelson opened a dive shop with a partner named Mathias Espinosa. They named their business Scuba Iguana Dive Center in honor of the reptiles he so dearly loves. The company is one of the most respected dive shops in the Galápagos and one of the most environmentally conscious. Part of Scuba Iguana's mission is to recognize "the interconnectedness and fragility of island life and the marine environment." The crew does coastal cleanups and remains active in campaigns to protect sharks, the very species most scuba divers come to see in the first place.

Recently, the company added two JetPak engines to one of its dive boats—powerful devices that use high-velocity pumps to eject water out the back end of the boat. The engines have several major advantages over outboards. The moving parts of the pump are located *inside* the engine, so there's no propeller to whack sea turtles or other marine creatures plying the coastal waters. Oil changes are minimal and the motors can survive up to ten times longer than four-cycle outboards.

Scuba Iguana offers diving classes ranging from beginning and open water PADI certification all the way up to dive master. Lately, however, running the internationally known company has been difficult. In 2007, without notification and for reasons no one fully understands, the Galápagos National Park closed down almost all live-aboard scuba operations in the islands. It allows only a few large dive ships to continue operating with legal licenses. For smaller companies the decision was a double-edged sword. On the one hand it created a bonanza for day-trippers like Scuba Iguana as divers flocked to their doors. On the other hand, as a travel agency Scuba Iguana makes most of its money selling dive tours on the live-aboard boats. "Financially, we took a real hit," said Nelson.

Bowing to unrelenting pressure, the park recently allowed a few more live-aboard dive boats, but they're all fully booked—well into 2010. It's frustrating for both Nelson and Espinosa. A pioneer dive master and naturalist guide who has logged more than 5,000 dives in the Galápagos, Espinosa is also a journalist and an underwater photographer. He was featured in the recent documentary, *Galápagos IMAX,* a film that focuses on marine biologists as they descend into the abyss in a submersible to explore the never-before-seen life-forms that thrive in the waters around these islands.

When Espinosa came to the Galápagos in 1987 from Quito, he went through the park's naturalist guide course, became a diver, then rose to the level of dive master. By 1991 scuba diving had really caught on, and soon a powerful dive company from the United States came to the Galápagos, the Aggressor Fleet, with two live-aboard yachts, the *Galápagos Aggressor I* and *II*. At the time Espinosa was one of the few dive instructors in the Galápagos. The Aggressor Fleet hired him as a dive guide and, eventually, as the chief of guides. But the severe shortage of professional dive masters forced the company to hire frogmen from the Ecuadorian Navy until Espinosa could train others. Many frogmen quit the Navy and began working for the Aggressor Fleet; it was much more fun, and the pay was like manna from heaven.

In Espinosa the company had found a real treasure; he is half German and speaks English and Spanish as well. As a trilingual dive master, he was invaluable for the new boom in Galápagos diving. Soon he advanced to master instructor and was teaching other dive instructors. He had reached the highest level possible and that's essential, says Nelson, who believes that finding the right dive teacher is like seeking a guru: "You go to a master who gets you started in this art because scuba diving really is not instinctive. You have to train to do things that aren't natural, especially in a panic situation." That's why he chose Espinosa as his business partner: He'd become a guru.

It's January 2008, and I'm eating lunch at La Garrapata, a popular Puerto Ayora restaurant. I look up to see a man dressed in the khaki uniform of a park guide. I recognize him as the man who spoke to my students at the Galápagos Academic Institute for the Arts and Sciences (GAIAS) in 2006: Mathias Espinosa of Scuba Iguana. He recognizes me, too, and I invite him to my table. Espinosa is a striking man with thick black hair, full lips, and eyes the color of sapphires. Like most park naturalists he's physically fit from a life spent outdoors and from diving in some of the strongest ocean currents on the planet.

Espinosa joined up with Nelson at Scuba Iguana in 1995 after a four-year stint with the Aggressor Fleet. At the time there still weren't enough skilled dive masters to meet the growing interest among foreign tourists. In fact, dive masters were in such demand around 1994, he says, that they were earning between $3,000 and $4,000 a day. "It spurred an economic refugee movement" from mainland Ecuador. But diving is so physically grueling that Espinosa knew only a small percentage of the new divers would make it. He wanted to involve the locals, especially those engaged in the illegal business of shark finning, sea cucumber harvesting, and the mutilation of sea lions by slicing off their penises, which could bring $50 apiece on the Asian aphrodisiac market. He wanted to show these guys the *real* Galápagos, but he needed candidates who had the right stamina, the ability to pass the tough diving exams (especially the math and physics sections), the desire to learn English, and the incentive to truly change their lifestyle.

"When someone offers you a way out of poverty, you have to bite the bone; you can't be lazy," he says. "Kids today are lazy." Espinosa searched all over Puerto Ayora for potential trainees. One day he found an orphan named Santiago, who was repairing bicycles. His mother had died of cancer, and a naturalist guide, a woman who was living in the Galápagos, adopted him. He'd been living alone since he was 12 and was trying to finish school.

"Santiago was a very independent character. He was a survivor. He worked very hard. I told him, 'Hey, Santiago, if you finish your schooling maybe there's a chance that you can find a job like a naturalist guide or something like that.' So he said, 'Okay.' "

"In 1994 or 1995 he came to me and said, 'Hey, here's the paper. I have finished school.' I said, 'Santiago, you know what? Let's organize a dive course for young fellows from here. I want to give the dive course. I want to teach you how to dive.' "

Espinosa taught Santiago and three other young men, one of whom, Edwin Naula, would eventually become the director of the Galápagos

National Park. It wasn't easy. "I'd ask them, 'So you want to be a dive master? Have you any dive license?' 'No.' 'Do you know how to dive?' 'No.' 'Do you know any other language?' 'No.' 'Have you worked before with tourists?' 'No.' " Eventually, the boys did learn. They viewed Espinosa as an older brother, a caretaker. Some worked with him for years.

One of Scuba Iguana's top divers was Juan Carlos Moncayo, known locally as Macarón, a former fisherman and shark finner. "He is an example of how somebody can change—somebody who was ready to sell the butt from his grandmother," says Espinosa. One night, though, something shifted deep within Moncayo's soul when his young daughter asked him why he killed sharks. The question rocked his world. Ashamed now, Moncayo soon joined Scuba Iguana as Espinosa's newest dive student. In time he became such a great dive master that he left Scuba Iguana and formed his own business, Maracón Tours, across from Pelican Bay.

Espinosa and Nelson agree that it's tough to invest so much time and money into someone's education only to watch him leave and become a competitor—in this case right down the road. "How do you know someone's heart?" Espinosa asks, thinking back. Even so, he and his protégé have remained friends with no hard feelings, and Scuba Iguana and Maracón Tours continue to train scores of Galápagos fishermen to become dive masters.

People move on, and in 2004 Espinosa, his biologist wife, María Augusta, and their two-year-old son moved to the sandy village of Villamil on Isabela Island. As Scuba Iguana's dive guru he had trained more than a dozen locals to become dive masters in this traditional fishing village. But good dive masters can survive only from six to eight years. Espinosa has been diving for 20 and his body, although still strong, was beginning to feel it. The frigid waters of the deep have caused chronic sinusitis and sensitivity to cold. That's why he made his last dive in

2007. Besides, he adds, the profession has changed too drastically since he first began diving.

"The work of a dive master 20 years ago was much easier than now because [those] who came, in general, were a good level of diver. Now [for us] it is a mixture between dive master and underwater babysitter. The level of divers who are coming now is very low. We have more and more people who want to go to the famous places [like Wolf and Darwin Islands] to see hammerhead sharks, in strong currents, and then you ask them, 'Okay, how many dives do you have?' "

" 'Twenty.' "

In the strong Galápagos currents even someone with 400 dives could be swept right out into the blue and never be seen again.

Not long after our meeting at La Garrapata, I take a speedboat to Villamil to talk with Espinosa and another famous dive master named Jimmy Iglesias. From my hotel I walk barefoot down a sandy road one block to Espinosa's modest home on the beach. We all sit on a cozy sofa to talk.

Iglesias was one of the very first divers in the Galápagos, a guy gutsy enough to learn the basics on his own. He did his first dive when he was 12 years old—solo. Iglesias loved the way the fan coral swayed like living rainbows and the sensuality of water on his skin. "It was the one material in life that could mold everything you are." Iglesias became a naturalist guide for the Galápagos National Park in 1981. At that time he says spiny lobsters were everywhere.

"How many did you see underwater?" asks Espinosa, who is sitting across from him.

"Up to 25 on a single dive."

"How many do you see now?"

"Almost none."

That was around the time the lobster industry crashed from over-fishing in the 1980s, and scuba diving was on the increase. Iglesias says guiding was really tough in those early days. "About 80 to 90 percent of the divers were beginners, and they would hide their lack of experience to not be intimidated by the others." At depths of 50 or 60 feet, he says, people would tend to get disoriented and panic. They'd grab onto him and become entangled with him and his diving gear, jeopardizing his life. So he invented something he calls his "magic broom." When an inexperienced diver panicked Iglesias would hold out his magic broom, make the diver grab it, then use it as a tow line, rising slowly up to a safe depth. "It was to save my life and theirs, too."

Espinosa first met Iglesias in 1988. Both men lived in Puerto Ayora and both were musicians. Espinosa played the guitar and his new friend played percussion. When they ended up as guides on the same dive boats, they'd entertain tourists by jamming. "It was like playing the blues in blue waters," Iglesias quips. Most passengers back then were adventurers from the United States, and the islands they wanted to visit were Darwin and Wolf, where the largest populations of hammerheads dwell. Once in 1989 when Iglesias was guiding a tour just north of those islands he came upon a Costa Rican fishing boat using a longline to catch sharks. Iglesias is a no-nonsense guy, tough as a Marine. He even looks like one. Enraged, he pulled out his video camera and captured the event on film. As the poaching vessel sped away its crew opened fire. "They were shooting small guns but we kept filming." Iglesias turned the film over to the park service, which passed it on to the Ministry of Tourism. It was the first documentation of illegal fishing in the area, and Iglesias won a conservation award for his courage. "I don't think I really deserved that award," he says. "I was just doing what had to be done."

The next year he filmed another illegal operation in Shark Bay off Wolf Island—hundreds of sharks caught in a gill net. The footage was shown all over the world and helped raise awareness of what was happening in the Galápagos. I ask the two men if they think the park is doing an adequate job of protecting the resources. Yes, they say, on the land on remote islands, but Espinosa adds: "I think that in the last few years the park is losing [perspective] on who are the bad guys and who are the good guys." As co-owner of Scuba Iguana, the second oldest dive shop in the islands and one of the most trusted, he says the company has always worked within the legal parameters established by the park. But the park service will not issue permits for small dive boats. Because diving has occurred here for nearly 30 years without any formal regulations, the activity is neither legal nor illegal. One has to question whether the new approach to issuing dive permits makes sense.

"If the park is moving as slowly as the giant land tortoises [to regulate] that activity, it's not our fault," says Espinosa. "It's strange that we are a small dive operation, and yet we're getting a lot of harassment from the park because we don't have a final permit. Why are they punishing us, the small operations, and not the big ones?"

"I don't understand it," says Iglesias.

Espinosa responds: "My impression is that the park is going with the current of human pressures. We have a big population of fishermen. They are saying that our tourism and diving activities are illegal. The fishermen group is very powerful, very strong. So they're saying to the park, 'Hey, this is illegal. If you don't stop it you're going to have problems with us.'"

He's right. Fishermen are now moving into tourism and getting most of the permits while the real pioneers—the brave men I'm now sitting with, guys who risked their lives using heavy steel tanks instead of aluminum—are being locked out. "Imagine," says Espinosa. "If the plumbers of Galápagos came and said, 'Okay, we want to be naturalist guides.' And the park says, 'Okay, you can do that, and you—Jimmy and Mathias— you cannot work anymore as naturalist guides because we need these jobs for the plumbers.'"

"Mathias," says Iglesias. "Have you gone to the park yourself and talked to the main guys about your point of view, and they still say, 'Sorry, we don't want to make any exceptions.' Is that what they say?"

"Yeah, we got screwed."

"Oh my God. I didn't know that!"

To get a permit, Espinosa explains, he'd have to partner with a fisherman, someone who's likely to say, " 'Hey Jack and Mathias, I'll rent you my permit for $5,000 per month. That's my bank account and I'll just deposit the money at the end of each month.' Fishermen are selling their permits under the table to other businesspeople. That's the real situation, man."

The next morning I return to Espinosa's house to meet with another long-time naturalist guide, Minino Bolona, a former soccer player from Guayaquil who rose to the national level at the age of 15. But the soccer hero had other plans, and in 1975 Bolona moved to Puerto Ayora on Santa Cruz Island to become a naturalist guide at a time when only about 500 people lived there. "Santa Cruz was so charming in the beginning," he fondly recalls. "We had a beautiful beach with white coral sand. But they stole the beach to build houses. Santa Cruz grew up very fast. It was my first love, but everything changed. It's not my Galápagos anymore." That's why he and his family eventually moved to Villamil, like other guides who got fed up with illegal immigration, urban sprawl, and chaotic enforcement of the laws.

Bolona and Espinosa met in Puerto Ayora at the end of the 1980s through a mutual friend. All three were musicians, and all were composing music about the wildlife and people of the Galápagos. In 1997 Bolona, who plays guitar, harmonica, and who sings the blues, won a first place award for a song he wrote about the islands. The prize: a certificate, a trophy, and a bottle of rum. Espinosa has cut a couple of

CDs, including one called *El Ultimo Paraíso (The Last Paradise)*. One of the songs was written by his wife, who also sings and plays the guitar. As a newcomer to Villamil, she was mortified every time she saw an unleashed dog chomp into one of the marine iguanas that lived right outside her door.

The intention of their music, Espinosa says, is "to get to the heart of the people and make them think about what a beautiful treasure we have, that we are very lucky to live here, and maybe convince them to pull the hair on their arms and wake up."

He's referring to the direct failure of Ecuador's educational system, especially in this province 600 miles away from the mainland. "The people who live here are not the grandchildren of Charles Darwin," he declares. "They are people who came to look for a better [standard of] living as farmers, and later fishermen. If we analyze why people came to Galápagos, I would say 99.9 percent came for economic opportunities and 0.0000001 percent came to live in the last paradise." With so many problems on mainland Ecuador, education in the Galápagos is one of the government's lowest priorities. The current educational system doesn't give students a chance to think, to be creative, to interact with the teachers. "Memorize it. Don't think," he says. "We have a state school system that is from the time of the dinosaurs, and nobody wants to do much about it."

Furthermore, says Espinosa, many schools in the Galápagos and on the mainland are in the hands of the church, "and the church says Darwin is wrong. It's creationism and not evolution that they're teaching." About five years ago he accompanied a camera crew from the BBC around Puerta Ayora to ask residents if they believed in evolution. "We interviewed like 30 people and we only found one or two who really believed in Darwinism. Then we got an invitation from the *evangelistas* to attend a church festivity and we filmed how all the animals were saved on Noah's ark. Here we are in the cradle of evolution, in the Galápagos, but nobody believes in it."

These paradoxes frustrate Espinosa. "We humans are egoists who have greedy thoughts. It's part of our genetic behavior all over the world. But the same people who are saying we need more and more are telling us, 'You need to slow down and if it's possible you could *leave* Galápagos because nature needs to be saved.' That's a gigantic contradiction, the size of the galaxy."

Meanwhile, he continues to produce musical CDs and documentaries about the Galápagos, including one on history that includes the still living pioneers, some of whom are now in their 90s. "One of the reasons I'm trying to teach to the local community to be dive guides and land guides here on Isabela is because I want to convince my species, *Homo sapiens,* to try to understand that what we have in Galápagos is the last chance to find a way to live in harmony with nature.

"If we cannot make that step in this last beautiful paradise on Earth, to find a harmonic way to live and use the resources in a wise way, I think it will be the last battle of the last fortress of nature," he says, his blue eyes brimming.

I want to know how the fishermen in the Galápagos became so powerful, so I track down a source who asks to remain anonymous. Let's call him Guillermo. "If you really want to know what happened," he says, "you need to know how one man changed the course of history in the Galápagos. He was a charismatic psychopath named Eduardo Véliz Riñones." Guillermo once worked on a boat with Véliz. "He was about 17 years old," Guillermo recalls, and had almost no experience whatsoever. "He was arrogant and difficult from the start, ordering everyone around," including men who had 20 or 30 years of experience. He even pulled a knife on one of them but was subdued. Véliz was determined to rise to the top. He worked as a naturalist guide and eventually became the sole congressman for Galápagos Province, from 1994 to 1996.

When the government tried to place a ban on sea cucumber fishing in 1995, Véliz went ballistic. He bullied the National Congress, thus politicizing the Galápagos and the Galápagos National Park. Véliz wrote his own "conservation bill" and pushed it through Congress. The bill demanded that the government lift the ban on sea cucumber fishing, give the locals more control over politics and the park, dismiss the park superintendent, and require that tourists stay at least one night on the islands, which would have led to massive and unplanned construction. But President Sixto Duran Ballen nixed the bill at the last moment and presented his own legislation. A few days later Congressman Véliz stirred a rebellion in the Galápagos, where he rallied the support of masked fishermen armed with machetes, bats, and axes. Riots ensued throughout the islands as the insurrection spread. Officials seized about 80,000 sea cucumbers that had been fished illegally and incinerated them. Eight of a group of about 30 fishermen were arrested; the rest fled.

Véliz had stacked the deck in favor of the fishing sector. Over the years he brought in hundreds of *comerciantes,* so-called fishermen, to support his cause. By 1997 the once small fishing population of about a hundred longtime locals exploded to over a thousand—most of them from the mainland. By that time, Véliz was persona non grata in the Galápagos. The sea cucumber population had been so overexploited by these newcomers that many fishermen were now risking their lives—diving too deep and getting decompression sickness—or the bends from nitrogen bubbles trapped in the blood. To many divers this was a welcome risk; the income they earned from middlemen in the Asian sea cucumber market was just too alluring.

Gabriel Idrovo is the medical doctor who saves the lives of such divers in the islands' only decompression chamber. The first thing you notice in his tidy office is a handmade sign that reads: *Primo non nocere.* It's the oath that enjoins this osteopath to "above all do no harm." It's also a philosophy of nature that Idrovo has adhered to since he first came to the Galápagos in the late 1980s. The way I see it, it's also a warning to the thousands of scuba divers who come to the Galápagos each year to see sharks and whales up close. Idrovo, who is also a dive master, runs the decompression chamber for scuba divers and has saved hundreds of lives. The recompression, or parabolic, chamber is part of a worldwide network of such facilities, and the only one in the Galápagos. The breathing mixture commonly used by scuba divers is a combination of oxygen and nitrogen. The percentages vary with different depths, as well as the anticipated length of a dive. At depths greater than 130 feet, the nitrogen component builds up in the diver's circulation. A rapid return to the surface does not allow enough time for nitrogen bubbles that have formed in the blood to dissipate. The result is nitrogen poisoning, and the pain is so severe that the diver will cramp up, causing him to bend over—thus the nickname, "the bends." If a diver makes stops during the ascent from a deep dive, the nitrogen bubbles decompress, and the blood returns to normal. Without such stops, it is necessary to enter a decompression chamber, which over time has the same effect.

On a cloudy afternoon after the rains have passed, Idrovo agrees to meet me and take me to the chamber. As I sit on his office steps awaiting his arrival, I watch two lava lizards mating on a boulder. The male's tiny red clasper penetrates the female so quickly I almost miss it. The two separate and run off as though nothing has happened. Idrovo, a compact man with kind brown eyes and a brush cut, arrives in loose green hospital scrubs and flip-flops. We walk up the road to the chamber a few blocks away and enter a nondescript building. Inside, though, the unit looks like something designed for space travel. It has windows made of Plexiglas, and access in and out is through special hatches and airlocks.

When the chamber was built in 2001, the goal was to treat both recreational divers and sea cucumber fishermen. The facility was long overdue. Even before the exterior walls were erected to house the unit, Idrovo and his staff were treating between 15 and 20 fishermen a week. In 2006 one diver descended to the deadly depth of 165 feet. "He was really in bad shape," Idrovo recalls. "It's challenging for us as doctors to treat this kind of patient because they arrive here in shock, paralyzed, dying." The 23-year-old fisherman recovered, though, and went right back to diving. "Crazy guy," says Idrovo, shaking his head.

Idrovo has treated well over 100 fishermen, about 40 tourists, and between 20 and 25 dive guides and instructors, the latter because they must dive more often and assist at-risk novices, which confirms the value of Jimmy Iglesias's magic broom. Idrovo has also rescued many divers, including those who get trapped in strong currents. Once, at a world-famous dive site called Gordon Rocks, he rescued a man caught in a swirl. The diver's eyes were bulging out of his head and spinning around like marbles. "I saw the depth meter on the computer," says Idrovo. "It said 140 feet." But he saved the man and later received a book about sharks as a thanks-for-saving-my-life gift.

Like most divers, Idrovo himself has been trapped in some odd situations. When guiding others he used to dive with a buoy attached to a line when he was leading a group. One day he suddenly descended about 164 feet in about ten seconds. A huge male sea lion had grabbed him, and by the time the bull let go, the line reel was almost empty. The good doctor had descended hundreds of feet! "I came up doing stops" to avoid getting bent, he says, his eyes wide with his recollection of the near-death experience. Another time near Darwin Island he had a brush with a huge toothed whale, an old leviathan that peered right into his mask. "I moved slowly to the side and he went with me. It was a fantastic experience." Once, near Lobos Island, two killer whales, a male and female, swam about ten feet away from Idrovo and his divers.

"They were just passing by, cruising. Two huge whales. Experiences like that keep me coming back."

On some dive tours Idrovo's wife, Ann, a botanist at the Charles Darwin Research Station, accompanies him. Back in 1988 and 1989 he was earning $30 a month as ship's doctor. Then he learned he could make $300 a month as a translator. Finally something kicked in. The naturalist guides were making $1,500 a month with very little training—ten times more than a multilingual medical doctor. He soon became a guide and worked for awhile on an infamous boat named *Bucanero*. "It was one of the worst," he recalls. "The captains were always drunk, and they had engine trouble. They'd hit rocks, other boats, even military boats." Idrovo quit. He and his wife returned to France for eight years while he advanced his medical skills, but the two eventually returned to the Galápagos.

When they arrived, the safety issue hadn't changed much at all. One of the worst things Idrovo ever witnessed was a boat, the *Bartolomé,* that caught fire when its metal chimney made contact with the wooden frame. Oddly, the boat was anchored just offshore the island of the same name. Idrovo, who was supposed to be guiding that boat but who got switched at the last moment, was actually on the island. Looking down he saw what was happening. The naturalist on board, a friend, was waving and yelling up to him as the boat went up in flames. One passenger appeared on deck, a ghastly ball of fire, before collapsing. Idrovo raced down to render first aid. In all, six people died, five guests and the cook. The cook had jumped overboard in terror and drowned. He didn't know how to swim. This tragedy is one reason guides are now required to take safety courses.

One evening I attend one of Idrovo's classes at a dive shop in Puerto Ayora where 14 fishermen are in training to become dive masters. When Idrovo gets to the section of his PowerPoint presentation on burns, several men nearly get sick when photos of *fourth*-degree burns come up on the screen.

Later I ask Idrovo what the biggest changes are that he's seen in the Galápagos since 1988. "The population," he replies without pause. "The increase from about 10,000 people to 35,000. There were perhaps ten cars when I came here. Now it's about 350 [mostly taxis] and they keep coming. The social pressure is incredible." And while the number of boats is supposed to have limits, new ones show up all the time.

"The main problem here is tourism. Everything happens because of tourism. There's money so people want to come here from the mainland. So what's the problem? People don't have the conscience that Galápagos is a very special place and they have to act differently here; it's not like on the mainland where you can have a dog, or a television, or a car."

Idrovo agrees with everyone I've spoken with that lack of education is destroying the islands. "I think the only people who can save the Galápagos are the children, because the adults' objectives are already embedded. If you train children they can think differently, love the place more, and learn how to protect it."

He's begun to see a change in his own young son and daughter in the way they experience the world. "The fact of being here has really given them a consciousness of the environment, how fragile it is, and that we have to enjoy it but also be part of the solution."

CHAPTER 8

Life on the Edge

"When one tugs at a single thing in nature,
he finds it attached to the rest of the world."

—John Muir

It's September 2005, my first trip back to the Galápagos since 1990. I've come to the islands to conduct research for a class I'm teaching at the Galápagos Academic Institute for the Arts and Sciences (GAIAS) on San Cristóbal Island. The tour company I've chosen is Ecoventura, which is said to be one of the greener tour operators in the archipelago. It's the beginning of the dry season, when the air and water temperatures are much cooler. I arise early, look out the tiny porthole in my cabin, and climb onto the deck. The morning is clear and calm. Blue-footed boobies drop from the sky like rockets into the surf for fish.

As tourists, we number 20, manageable when broken into two groups of 10. After a breakfast of eggs, bread, yogurt, and tropical fruit, we're ready to visit one of my favorite islands: North Seymour. We don life jackets, separate into two dinghies, and motor a short distance to the island, one of the best bird-watching sites in the Galápagos. Sea lion pups peek out of the water as we pass, then disappear in shafts of bubbles. We dock near a colony of marine iguanas, and as we climb the trail, swallow-tailed gulls stare from pitch-colored rocks with eyes ringed in scarlet. Though I've been here before, it still seems extraterrestrial, like a faraway planet.

A new generation of boobies has begun to hatch, and fuzzy white chicks peer from beneath their mothers' breasts. Others look way too big, in that awkward stage between hatchling and juvenile. As we follow a clearly marked trail through prickly pear cactuses we must watch where we step: Boobies build their nests on the ground. There are some with feet the color of sky and others with feet the color of red lipstick. But the strangest species, the Nazca booby, looks like it is peering through a chocolate-colored mask, like a character out of Dr. Seuss. The rookery is a symphony of songs and squawks and the flutter of wings—the epicenter of a mating frenzy unparalleled in these islands. I stop to observe a red-footed booby on a nest of twigs, preening its chick. A few feet away a young blue-footed booby tests its wings, propelling itself in short little spurts from one boulder to another. Male frigate birds soar overhead, their throat pouches inflated like big red balloons as they try to attract females. Their cries sound like door hinges in need of oil. In this frenzied state it's a wonder they can fly at all.

Our guide on this seven-day trip is Edwin, a naturalist who has worked for the Galápagos National Park for more than a decade. As we stop in the shade of a dwarf *palo santo* tree, he recounts the time he saw four male frigate birds perched in a single tree, vying for a mate. As the jealousy intensified, one male pecked at his rival's pouch until it fizzled like a punctured balloon. A death blow from a sharp hooked beak. If a male succeeds in attracting a female, he brings her twigs to begin building their nest. She must do so carefully, for if the single egg falls out of the nest he'll leave her. "A divorce," Edwin says, slashing his finger across his throat.

Edwin is short and stout, with calves as hard as mahogany. His hair is close-cropped and two days of stubble pokes from his chin. Lately he's seen changes in the land—changes that cannot be ignored. Beyond the rookery, the feathers and skeletons of baby frigate birds poke eerily out of the bushes. The chicks, he tells us, were abandoned by their parents, who had to fend for themselves or starve. "How do you know?" I ask. "Lack of cloud cover," he says, pointing to the dried flipper of a sea lion half-buried in the sand. Then he shows us the decomposing carcass of a booby chick and succulent plants that should be green by now. The rains have been scant this year, and Edwin predicts that another El Niño cycle is on the way. If he's right, marine life could suffer drastically. The last two El Niños, in the 1980s and 1990s, warmed the ocean so much that the Galápagos penguin population was cut nearly in half. A fine balance of phytoplankton is necessary for a healthy ocean environment. Warm water in this nutrient-rich region creates too much plankton. The result can be huge algal blooms that deplete oxygen and can suffocate marine life.

Along the trail we come upon two male land iguanas that have squared off in battle beneath a patch of prickly pear cactuses—one of their favorite foods. Suddenly, one iguana raises his tail and thrashes the other. *Thwap!* The reptile hisses, repositions his body, and strikes back. The blow seems more like a warning than an agressive move, and soon the loser backs down and waddles across the path, right in front of us. The confrontation is too fantastic to seem real. We witness this spectacle as if we're invisible. There's no instinctual fear in these animals whatsoever, a trait that often astonishes visitors to the Galápagos. In 1835 when Darwin set foot on the islands he wrote: "A gun here is almost superfluous, for with the muzzle of one I pushed a hawk off the branch of a tree." Of the other terrestrial birds, he wrote: "There is not one which will not approach sufficiently near to be killed with a switch, and sometimes, as I have myself tried, with a cap or a hat."

Science writer David Quammen states in his book *The Song of the Dodo: Island Biogeography in an Age of Extinction* that "tameness" is the wrong description. "These animals aren't imbecilic. Evolution has merely prepared them for life in a little world that is simpler and more innocent than the big world."

Nothing could show this to be more accurate than the flightless cormorants we see later on Fernandina Island. This is the only cormorant species in the world that has lost the ability to fly. As we pause near the rookery, a few of these odd-looking birds emerge from the waves. Like mainland cormorants, they stand with "wings" outstretched, drying their stunted appendages in the sun as though some genetic memory has kicked in. They've been diving for squid, eel, and fish about 300 feet offshore. Now they waddle back to their mates to present their catch and exchange places on the nest. A soft grunting fills this colony at the edge of the sea, as though there is exquisite perfection in the swapping of parental duties on nests made from algae, sea urchins, and sea stars. When I squat down for a closer look a newly hatched chick pokes its featherless head out from beneath its mother. The newborn resembles a tiny gray snake, and all the while the mother seems oblivious to my presence, her steely blue eyes staring off into space.

"Why have these cormorants lost the ability to fly?" I ask Edwin.

"Everything they need to survive can be found all around them," he responds. "Eventually, their wings will either fall off or become flippers, just like penguins. Use it or lose it."

They look like mutants, a cruel joke, an evolutionary mistake. Yet, standing here on a carpet of red and purple sea urchin spines, I'm struck by what seems a perfect example of adaptation: evolution unfolding right before my eyes. I want to know more about the species, and in 2008, I would have that opportunity.

Carlos Valle, one of the world's leading experts on the flightless cormorant, is a professor of ecology and evolutionary biology at the Universidad San Francisco de Quito (USFQ) and co-director of GAIAS in Quito. There, in the summer of 2008, we meet at an upscale restaurant between our respective classes at USFQ to discuss his research. Valle is a small man with dark wavy hair and whimsical eyes that peer through round glasses. We order fresh salmon from Chile with avocado and finely chopped chilies.

When Valle was 19, he worked with scientists as a field assistant to count penguins and flightless cormorants around Fernandina and Isabela Islands. His undergraduate thesis focused on cleptoparasitic behavior among great frigate birds: the habit of stealing food from other sea birds in midair. Administrators at the Charles Darwin Foundation (CDF) were so impressed with Valle they soon appointed him an associate scientist to do census work on penguins, cormorants, and flamingos, among other birds. Valle later worked as a naturalist guide in the Galápagos for a couple of years, and then moved to the United States to advance his education. He was accepted into Princeton University and was assigned an adviser by the name of Peter Grant. Grant and his wife, Rosemary, are the world's leading experts on Darwin's finches. "They were just wonderful," Valle recalls.

The young naturalist-ornithologist wanted, eventually, to return to his home in the Galápagos, where he was born. He'd become interested in the mating habits of the flightless cormorant and was fascinated by the behavior of the female. The female cormorant helps incubate the egg and feeds the hatchling for only two months before she deserts the nest—and her mate. She simply wanders off, leaving him to do the parenting for four to nine months while she mates with others. For Valle, who spent years in the field, the myth of monogamy in the cormorant was blown right out of the water. "Cormorants are an exception to the rule," he explains. "Sexual selection doesn't fit the other patterns in the Galápagos."

Biologists call this behavior of the female cormorant sequential, or serial, polyandry. Valle says the process raises interesting questions from an evolutionary point of view, because the courtship process is reversed. "With birds, who does all the dancing around in courtship?" he asks me. "It's usually the males," he says, answering his own question. "Males fight. Males dance to get the attention of the females. In the flightless cormorant that is not the case. Females do the fighting to get closer to the potential mate." In fact, six or seven females might dance around a single male, fighting for supremacy. The male, on the other hand, does what female birds usually do. He chooses his mate, and she's often the strongest. I'm beginning to grasp Valle's obsession.

"But," I ask, "why have they lost the ability to fly?" Valle sets down his fork and takes a sip of wine. He's in his element now. In fact, he admits he's doing research to investigate some new theories he's recently come up with. One can only speculate about the evolution of flightlessness. For one thing, the cormorant has no terrestrial mammalian predators, so it doesn't need wings to escape. "They're bottom-feeders and slow swimmers. They only need to swim about 200 meters from the shore." And because they can't fly, they can avoid those pesky frigate birds, who steal fish from other birds in midair. "Flying is a very costly trait," he concludes. This is something that never occurred to me. I've always thought that having wings and soaring above the heavy gravity of Earth was the perfect symbol of freedom.

Then Valle stuns me by suggesting that the flightless cormorant might be inbred. "So they're mutants?" "They're mutants, exactly." They still need their wing feathers to stay warm in the cold waters of the archipelago, and some hypotheses claim they still air out their "wings" to avoid getting parasites or algae growth. And, he suggests, by spreading its wings, the cormorant is extending its thoracic box to help swallow fish. I ponder his marvelous theories. Could the cormorant's wings become vestigial appendages like long-lost tails? Will this idiosyncratic bird jump species and become something entirely different? If the fossil record is true—that the modern whale some 52 million years ago was a land mammal the size of a wolf that gave up its legs for the sea—I'd say anything's possible.

Who knows what's true? One truth is that the isolation of the Galápagos Islands is exactly what makes their life-forms so exotic. But therein lies the trap. No one could have anticipated how radically things would change or how rapidly some species would become threatened or extinct once humans arrived on the scene. One of the keys to understanding the islands is that they're no longer isolated. The ease of getting to the archipelago has created a vacuum to be filled by colonists, tourists, and invasive species. Take for example the Galápagos penguin, an iconic cold-weather species that looks implausible on the scorching equatorial islands. It's tiny, only about 20 inches high, and it weighs less than a watermelon. The International Union for Conservation of Nature (IUCN) listed the species as endangered in 2006 and estimates that only about 2,100 Galápagos penguins exist today. Until recently, the population was much higher, but El Niño of 1997–98 decimated it by 65 percent. There are also unnatural threats: Penguins are easy prey for illegal fishing, oil spills, invasive species like mosquitoes, and garbage dumped into the sea by fishing and tour boats. The CDF and the Galápagos National Park, in a recent collaborative study, predicted that the Galápagos penguin faces a 30 percent chance of extinction in the next hundred years from global warming alone. Human threats could hasten their demise.

Like other birds of the islands, the mangrove finch is now critically endangered. This crafty bird feeds mostly on insects. It has learned how to pry insects or larvae out of wood, using a cactus spine or a twig poised in its specialized beak. Mangrove finches are now extinct on Fernandina Island, where the flightless cormorants live. On nearby Isabela Island the mangrove finch is threatened by habitat destruction and invasive species, including cats, black rats, fire ants, fly larvae, and the ani, an aggressive member of the cuckoo family. Diseases such as avian pox make the finches of the Galápagos even more vulnerable. Scientists at the CDF are monitoring and banding mangrove finches and taking blood samples. They now run a captive-breeding program to release the birds into the wild and thus prevent their extinction.

Land iguanas were once so abundant on Santiago Island that in 1835 Charles Darwin couldn't find a place to pitch his tent. Today, humans and other invasive species have forced them into extinction on that very same island. They're also extinct on Santa Fe Island. Typically, the iguana lives in the arid lowlands and feeds mostly on prickly pear cactuses, although a tasty centipede would also be welcome. These noble reptiles can live for 60 years. Females can lay up to 20 eggs, but the hatchlings may fall prey to Galápagos hawks, egrets, and herons. Land iguanas sometimes mate with *marine* iguanas on South Plaza Island, where the territories of the two species merge. The result is a hybrid iguana with features from both species that's believed to be sterile. Like other endemic species, the land iguana is vulnerable to feral dogs and cats. Also, goats compete for vegetation the iguana depends on, and pigs dig up the nests to eat the eggs. On Baltra Island, where the United States built an air base during World War II and whose main airport is today located, the iguanas became extinct by 1954. The CDF continues a captive-breeding program for land iguanas to repatriate them to their original habitats on Baltra, Isabela, and Santa Cruz Islands.

Among marine life the sea cucumber, *Isostichopus fuscus,* is one of the most depleted species in the Galápagos. It is related to sea stars and sea urchins, and it plays a major role in the marine ecosystem by recycling nutrients that benefit other species. The sea cucumber is a sausage-shaped animal that lives in deep waters or on the seafloor. *I. fuscus* is especially vulnerable to humans. In Asia this creature is considered a delicacy—and an aphrodisiac—and it is heavily traded on the black market. The commercial exploitation of the sea cucumber began in the Galápagos in 1993. Then, in 1999, when the sea cucumber season legally opened, local fishermen raced to share in this newfound bonanza, ignoring legally set quotas. Between six million and ten million sea cucumbers were harvested in only three months, and this trend continued until *I. fuscus* was almost fished out. Scientists at the CDF are studying the species by growing it in aquariums and comparing the captive animals to those that survive on the ocean floor.

Twenty-seven species of shark are native to the Galápagos, including whitetip reef sharks, whale sharks, and tiger sharks. The presence of this ancient creature is one of the main reasons why scuba divers from all over the world come to the islands. Where else can you swim with hundreds of hammerheads and return home to talk about it? Sharks face a major threat from fishermen, but we'll get to that story later.

And what about the endemic species, which are found nowhere else on the planet? Why do they matter? Consider scalesia, for example. It's a member of the daisy family that evolved into a giant tree, like the magical beanstalk that sprouted from a seed that a mythical boy named Jack once planted. Scientists believe that scalesia descended from a single ancestor that reached the Galápagos long ago. The plant was so successful at adaptive radiation—the ability to survive and adapt in a distinct ecological niche over time—that today at least 15 species and 5 subspecies are endemic to the Galápagos. Later on Isabela Island, I would talk with one of the world's experts on scalesia—an 84-year-old former priest who raises the plant in his organic garden and distributes it to the same local ranchers who once plowed it under. The CDF is monitoring threats to this unusual plant, including agricultural expansion and invasive species. Among the latter are Cuban cedar, guava, red quinine, blackberries, passionfruit, and elephant grass, which can choke scalesia out of existence. The Galápagos is home to about 500 native plant species, more than a third endemic. Six hundred other plants are invasive.

Many foreign insect species have also entered the Galápagos biotope, like the moths that follow the lights of tour boats and threaten the biodiversity of the islands. The CDF reports that over a three-hour period in 2007 during the dry season, scientists counted an average of

150 insects around a tour boat that had 18 external lights. During the rainy season that number tripled. The worst vectors, say the foundation scientists, are international cruise liners such as the M/V *Discovery*, the largest ship that visits the Galápagos.

———————

The truth is that the Galápagos Islands are a bioregion with more endemic species than any other archipelago in the world. That's what makes the islands special and so sensitive to tourism. It's also one reason why no bridge has ever been built across the channel from the busy airport on Baltra Island to Santa Cruz Island—the main tourist hub. The threat of invasive species is just too risky, even from island to island. When I lived in the Galápagos in 2008 several new pests arrived, including the Mediterranean fruit fly (the medfly), probably from fresh produce shipped from the mainland. Around the same time a marine biologist discovered in Galápagos penguins a blood-borne parasite carried by mosquitoes called *Plasmodium* that can cause avian malaria. (It wasn't yet clear at the time which type of *Plasmodium* had affected the penguins.) Researchers are studying more individuals to identify the parasite, determine which mosquito transmitted it, and evaluate its presence in other bird species. *Plasmodium* species that infect birds, they say, cannot be transmitted to humans.

———————

And what of *Homo sapiens* and our own impacts on the parts of the planet we call home, far away from the Enchanted Islands? Here are a few startling statistics that put things into global perspective. The world (led by China and followed by the United States) now produces about 30 billion metric tons of carbon dioxide per year. A metric ton equals 2,205 pounds, or the weight of an average car. The United States alone uses about a third of the world's nonrenewable energy. It also emits a quarter of the world's greenhouse gases—7.2 billion metric tons per year. That's about 40,000 pounds *per person* a year, or the equivalent of 25 polar bears, or the total weight of the bombs that U.S. warplanes dropped on Baghdad in a ten-minute airstrike in January 2008.

Global warming? Natural greenhouse gases like water vapor, carbon dioxide (CO_2), and methane are essential in the atmosphere to absorb and release solar radiation and maintain Earth's temperature beneath a protective layer of ozone. Without these gases, the planet would be too cold to inhabit. Sunlight warms the planet enough to make it livable, and the rest of the heat gets radiated back into the atmosphere. But human activities since the industrial age began in the 1700s have created dangerously high levels of greenhouse gases and have steadily depleted the ozone layer. Higher levels of greenhouse gases—mostly CO_2 from burning fossil fuels like gas, oil, and coal—trap solar radiation before it can escape back into space. Think of Earth as a mirror where the radiation gets bounced right back in your face. The result is global warming—a gradual rise in air and ocean temperatures. Other greenhouse gases include chlorofluorocarbons (CFCs) and nitrous oxide. The UN Intergovernmental Panel on Climate Change now warns that the world is destined for a hotter future than predicted. The panel reports that temperatures could increase by 4°F to 11°F by 2100. Even an increase between 3.2°F and 9.7°F could trigger massive environmental disasters and cause sea levels to rise by about 82 feet.

Now consider methane. Agriculture produces about 18 percent of the world's greenhouse gases. The UN Food and Agriculture Organization believes that number could increase by 60 percent by the year 2030. Most methane comes from burping, flatulent cows, pigs, and sheep. The average dairy cow can blast out more than a hundred gallons of methane a day. In 2003 the government of New Zealand proposed a flatulence tax, but ranchers put up such a stink that the idea was squelched.

In 2008 scientists officially adopted a new word for the current geological epoch: the Anthropocene. The term was coined in 2002 by Nobel Prize–winner Paul Crutzen, an atmospheric chemist who argues that humans are directly responsible for much of the planet's climatic trauma. Crutzen came up with the term at a conference when someone referred to the Holocene epoch as "current." The chemist pondered the remark and found it inaccurate: The world had changed too dramatically during the last 10,000 years. Crutzen responded, "No, we are in the Anthropocene." He later said he'd made up the word on the spur of the moment. "Everyone was shocked," he said, "but it seems to have stuck."

Anthropo is Greek for "human," and *cene* means "new." In other words, it's a new age of our own making. Consider the following: According to Conservation International, every 20 minutes a species is forced to the edge of extinction and more than a thousand acres of forest are destroyed. Today, fully 99.9 percent of all species that ever existed on Earth are extinct. The Earth has experienced numerous mass extinctions, the last—the Cretaceous–Tertiary—occurred about 65 million years ago when most terrestrial life was wiped out, including the dinosaurs. The Anthropocene is the sixth major mass extinction, and scientists agree that humans are mostly to blame. Why? Overpopulation. Natural and human-induced climate change. Mass consumption. Pollution. Invasive species. Disease-bearing vectors. Overharvesting of natural resources. Habitat destruction. Loss of biodiversity. Ignorance. Today, nearly a quarter of the world's mammalian species are at risk of extinction, according to the International Union for Conservation of Nature (IUCN). As E. O. Wilson so eloquently stated in his book *The Future of Life,* millions of species may vanish before they're ever discovered.

Take for example a "new" bacterium found in October 2008 in a gold mine near Johannesburg, South Africa. The bug lives 1.74 miles deep in the earth at 140°F, surviving without oxygen or light. It relies instead on water, hydrogen, and sulfate. The bacterium is believed to be the first known organism to act as its very own ecosystem. This is astonishing, say scientists. If life can exist on Earth without oxygen, could it survive on other planets? Is this a clue to the origin of species? What would Darwin say? The team dubbed this strange new entity *Desulforudis audaxviator*. The species name, "bold traveler," comes from a Latin phrase in the Jules Verne novel *Journey to the Center of the Earth*. *Descende, Audax viator, et terrestre centrum attinges*—Descend, Bold Traveler, and attain the center of the Earth.

At the same time the "bold traveler" was discovered, Ecuadorians approved by a wide margin a new constitution that will expand the powers of the president and could allow Rafael Correa to remain in that office for a decade. The new constitution promises more spending on health care for the poor, provides more rights for indigenous peoples, prohibits discrimination, respects private property, allows civil unions for gay couples, and provides social security to homemakers. But to many, the most unusual guarantee is the "Rights of Nature," which declares: "Nature or *Pachamama* [Mother Earth], has the right to exist, persist, maintain and regenerate its vital cycles, and its processes in evolution." It mandates that the government "restrict all the activities that can lead to the extinction of species, the destruction of ecosystems, or the permanent alteration of the natural cycles." While many people lauded this article of the proposed new Constitution as visionary, others condemned it as hyped-up "greenwashing."

A few days after voters accepted the new Constitution, an indigenous group organized as the Fundación para la Sobreviviencia del

Pueblo Cofán (Foundation for the Survival of the Cofan Community) issued an urgent press release. The Cofán and Kichwa, members of this indigenous culture in the Ecuadorian Amazon, stated that three barges, hauling oil rigs tractors, and trucks for the state-owned company Petroamazonas, had attempted to enter their territory to begin massive oil drilling. The land in question lies on the edge of the Cuyabeno Wildlife Reserve, a forest and wetland habitat located in a major watershed. Petroamazonas had come, without notice, to take over a site left by the California-based Occidental Petroleum Corporation. The Ecuadorian government had kicked Occidental out of the country in 2006, claiming the company had illegally transferred 40 percent of its stake in Ecuadorian projects to a Canadian oil company without government approval. When it is operating, the site in Cuyabeno produces about 95,000 barrels of oil a day.

The intrusion by Petroamazonas into their homeland caught the Cofán and Kichwa by complete surprise. Armed with wooden spears, tribal members and the neighboring communities launched a citizens' arrest. They cited the brand new Constitution and its guarantees to nature and indigenous cultures and demanded that the laws of Ecuador be honored. The Kichwa then seized the barges and other equipment until legal counsel could be sought. Not once, they claimed, had Petroamazonas consulted the indigenous communities, nor had the minister of the environment in Quito, who administers the reserve, said a word. (She also administers the Galápagos National Park.) The Cofán said government officials confirmed that Petroamazonas had not filed a formal environmental impact statement, and that the Ministry of the Environment had not issued environmental licenses for the operation.

As for the Galápagos, President Rafael Correa has said he wants the Galápagos National Park removed from UNESCO's List of World Heritage in Danger as soon as possible, by transforming it into a global example of pristine conservation. This seems unlikely, given the political game of musical chairs on the islands and the continuing issues that plunged this World Heritage site into crisis in the first place: uncontrolled tourism and immigration, illegal fishing, invasive species, and a clash of fundamental values.

CHAPTER 9

"Judas" Goats

"All animals are equal, but some animals are more equal than others."
—George Orwell

Victor Carrión is a compact man with dark hair and almond-shaped eyes. In his drab green jumpsuit, he reminds me of my father, an Air Force fighter pilot who served in three wars. Carrión and his men have also flown in battle over hostile terrain and scoured canyons on foot for the enemy: in this case, goats—four-legged ungulates that have invaded Isabela Island and laid low its residents, the giant land tortoises that have lived here for hundreds of thousands of years. On their mission, the men's arsenal included AR-15 semiautomatic rifles fired from helicopters; .223-caliber rifles for hunting on foot; one million rounds of ammunition, including exploding non-lead bullets; telescopic sights; two-way radios; GPS and GIS devices; dogs imported from New Zealand; and "Judas" goats to rally the wild herds.

On an overcast afternoon in July 2008 I visit Carrión in a small office at the Galápagos National Park headquarters. I want to learn more about his work as one of the co-directors of Project Isabela, one of the largest island restoration projects in the world. The $10 million eradication campaign began in 1998 after feral goats and other invasive mammals multiplied exponentially on several islands and competed with the endemic wildlife. The project, he says, is a joint effort by the

Galápagos National Park and the Charles Darwin Foundation (CDF) to eradicate invasive species and restore the islands to their original state. Carrión, who is the control and eradication chief for the park, put together a team of highly trained wardens. They and the CDF received funding from the United Nations and several other nonprofit organizations around the world. The project received the blessing of the then-minister of the environment in Quito, who reports directly to the president. Felipe Cruz, whom I'd spoken with earlier, worked as the technical assistance director for the CDF. Cruz rallied the best scientists possible and provided state-of-the-art technology for the ongoing project.

In a five-year test run, the hunting teams had already eradicated about 89,000 feral goats on Santiago Island (plus wild pigs and donkeys). Now that the sharpshooters had refined their skills they moved west to Isabela where more than 100,000 feral goats were competing for what remained of the once verdant habitat with endangered land tortoises. As the largest island, Isabela has the richest diversity of endemic plants and animals. Up in the lush northern region, Volcán Alcedo supports an estimated 5,000 tortoises, the single largest population anywhere in the Galápagos. But the goats had grazed the native grasses down to the roots, especially near the rim of the caldera, and something needed to be done. Fast. The wild goats had also demolished other foods the tortoise depends on: succulent cactuses, leaves, vines, fruit, ferns, and bromeliads. Once these too were gone, the goats stepped onto the backs of tortoises, using them as footstools to reach moss high up in the trees. Biologists knew that without the trees the venerable old tortoises would die: They depend on dew pools created when mist drips from moss-trees. Tortoises drink from the pools and wallow in them to cool off, and to ward off ticks. Now the goats were drinking the pools dry. Land tortoises here can survive more than a year without drinking: They break down their body fat to produce water. But as the pools dried up and the soil heated, the competition between species changed the microclimate in tortoise nesting areas as well. Soil temperature determines the sex ratio of hatchlings. If the nest temperature is too hot, the conditions produce all females; too cool, all males.

As the reptiles shriveled and starved or slipped on ground eroded by goats and fell to their deaths from the caldera's rim, scientists at the CDF and the Galápagos National Park debated how to eradicate the goats. The sister organizations teamed up and created a strategic plan that was so ambitious it seemed impossible to accomplish. They trained aerial sharpshooters to hang out the sides of helicopters and blast the goats away with the AR-15 rifles. Down on the ground, teams of hunters carrying .223-caliber rifles scoured the rugged terrain with dogs trained to ferret out goats and gather them into herds without harming them.

In the final stage of the assault, the men released about 700 sterilized "Judas" goats from New Zealand on strategic parts of the island. The Judas goat's moniker comes from its ability to betray its own species. Each goat was fitted with a radio telemetry collar attuned to special equipment in base camps and in the air. Goats are social animals that seek each other out. When a Judas goat joins a herd, its collar triggers a signal to the hunters that exposes the exact location. I can only imagine what happens next.

I ask Carrión if the park plans to change the goat's nickname now that the "Judas Gospel" has proven among biblical scholars that Judas Iscariot was not what he seemed. He was in fact a Gnostic, like Jesus, part of a mystical religion that predated Christianity. Judas, the gospel states, was Jesus' closest spiritual ally—an entity capable of releasing him from the prison of human form by exposing him to authorities.

Carrión responds: "We have already changed the word. We now call them *chivos delatores*."

"What does *delator* mean?" I ask, unfamiliar with the term.

"The same. A delator is one who reveals a secret, like an informer. But it's been very difficult to get that name to stick."

———————————

As the aerial hunt continued the goats adapted. They learned to identify the sound of rotors with danger and hid in shallow caves or in the shade of trees. Meanwhile, the carcasses piled up on the ground and subsistence hunters from southern Isabela protested. They accused the park of competing with them and of leaving the meat to rot in the field. Removing the carcasses, however, was not an option. In the tropical heat the bodies decompose within a week and Galápagos hawks, native beetles, and flies scavenge what's left. "The terrain is just too rugged. There are very few trails, and there's no fresh water," Carrión explains.

The solution: Project Isabela hired about 20 disgruntled hunters to work with the ground teams, where they traversed palo santo forests, sashayed through cat's-claw bushes, crossed deep crevices, and used machetes to hack through the dense foliage around Alcedo. But there were other challenges as well. International animal rights groups caused an uproar over hunting methods. "We started using smaller and faster bullets that are more humane and cause less suffering," Carrión explains. Soon afterward the activists backed off. Then in 2004 one of the helicopters crashed. No one was seriously injured, but the chopper died on the spot, a useless wreck.

"How did the sharpshooters avoid hitting the tortoises?" I ask.

"No tortoises were killed. It was really easy to shoot goats from the air," Carrión explains. "On foot it was much more difficult." He picks up his ballpoint pen and holds it at a 45-degree angle to simulate how far the helicopters tilted when the shooters fired out the windows. The choppers sometimes dipped to only 49 feet. "At that level it's easy to see if a tortoise is near a goat." The New Zealand pilots also worked as sharpshooters. "Those pilots were excellent," he says. "They were very expert—and very expensive." About 95 percent of the staff, though, were Galapagueños who'd been trained in reading topography, hunting techniques, GPS and GIS operations, dog handling, and medical emergencies. The best shooter was a young man named Wilson Cabrera from Santa Cruz Island, who had learned to hunt with his father when he was eight years old. Cabrera scored a record of 1.8 bullets per eradicated goat. He later said that Project Isabela had taught him a wise lesson about invasive species: that humans must adapt to the Galápagos and not the other way around.

Ecuador's goal is to make the Galápagos "goat free" by the year 2010. It's a Herculean task in a remote province with a history of turbulence. Carrión knows this all too well. In 2004 he was the director of the Galápagos National Park when Ecuador was in complete chaos. "To be park director at that time wasn't easy because of all the political instability on the mainland," he says. Carrión served under the military dictator, President Lúcio Gutiérrez Borbúa, who the same year declared a state of emergency and dismissed Ecuador's new Supreme Court. Not long afterward Gutiérrez was ousted from office and fled to Brazil, where he was granted asylum in 2005.

"I lasted exactly 46 days," Carrión says softly. I detect no bitterness in this highly respected park official. It's as though nothing ever comes as a surprise in the islands. Carrión was about the tenth director of the Galápagos National Park in less than *two* years, a perfect indicator of the discord and disconnect on the mainland. During his six-week tenure he'd created a draft management plan for the troubled park, but his new vision for the Galápagos sat on shelf collecting dust. He was forced to step down until peace could be reached between the central government, the park, the CDRS, and the Galápagos fishermen, who still wielded tremendous political power throughout the islands.

Outside Carrión's office rain beats down on the roof. He's proud of his work on Project Isabela, which is now regarded as one of the most successful island eradication programs in the world. "It's one more step for the restoration of Isabela," he says. But he and his crew aren't hanging up their rifles just yet. The next targets are Floreana, San Cristóbal, and Santa Cruz Islands. "By then we hope to have eradicated all the goats in the Galápagos."

Felipe Cruz, who partnered with Carrión on Project Isabela, works for the CDF. He admits that blasting away all those goats weighed on him, but he agrees with Carrión that the teams fulfilled their mission. "We can say with pride that we even surpassed what was hoped for," he told colleagues. "We have done what the world thought was impossible, making feral goats, pigs, and donkeys now a story for the history books about Santiago and northern Isabela Islands." Cruz later wrote that in conservation biology there's less talk now about saving a species from extinction. "Now it is a discussion as to *which* species can be saved. We are rather like a collective Noah, deciding with a biblical coldness which life-forms will be able to accompany us on our new journey in the ark."

Everyone agrees that Project Isabela could not have worked without man's best friend. The same dogs that toughened their paws on volcanic rocks and rounded up goats under the sizzling sun now live in kennels behind Carrión's office. I ask if I can see them. "Of course," he says, slipping into knee-high rubber boots. Carrión leads me outside where the rain has cooled the heavy air. When he opens the gate to the kennels the dogs go crazy. Our first stop is the cage of Buck, a yellow Labrador retriever donated by the nonprofit group Wild Aid and trained to sniff out illegal marine products such as shark fins. Buck looks sad in his chain-link cage with only a water bowl for company. I notice that one of his hips is off-kilter as he sits on the concrete floor of his cage, gazing out at us. Carrión says it's all right to pet him, so I poke my hand through the door, stroke the fur on his head, and speak to him in Spanish, but Buck shows no emotion. I know he's a working dog, but as the former owner of a yellow lab I want to bring him a blanket to sleep on. And a tennis ball. And maybe some dog biscuits. But I can't: It's against park rules. A second sniffer dog donated by Wild Aid—a black lab named Aggie—is away today. She's chained up at the airport on Baltra Island awaiting luggage from flights to and from the mainland. I've seen her there many times, panting in the heat.

Next stops: A mutt named Gus who fell into a *grieta*, or crevice, on Isabela but survived; a black dog with a white muzzle named Tyson; a female with pointed ears named Charles, who slurps from her water bowl; and another female named Santa. There's Pluto, Delta, Duke, and Conan, Oso, Hoki, and Pluma. A dog named Rufo announces his name when we pass: *Roof-roof!* Scott looks ferocious as he bites the metal fence that confines him. Caña leaps up and down like a pogo stick. Near the end of the lineup we stop at the cage of Jane, a small dog with black and white polka dots. "This one I really like," Carrión says. He opens the kennel, lets her out, and pets her. The other 42 dogs go wild in a cacophony of barks, howls, and yelps. Carrión tells me he has two Labrador retrievers at home. "They're great dogs, especially with children."

In May 2008 I'd witness just how effective Project Isabela had been during a grueling trek up Volcán Alcedo with wildlife photographer Tui de Roy, who grew up in the Galápagos. The former naturalist guide for the park has spent almost four decades documenting the giant tortoises on Alcedo, one of her favorite places on the planet. In an article for the Galápagos Conservancy in 2006, de Roy wrote: "When the goats began invading in the early 1980s I began witnessing the slow agony of the tortoises and their prehistoric wonderland." In 1995 she documented the loss of mossy fog-drip trees along the rim of the caldera, which the tortoises depend on. When de Roy returned to the denuded landscape in 2000 the tortoises were beginning to die from lack of shade. They were literally baking under the equatorial sun. Then in 2005, when Project Isabela was in full swing, de Roy traveled to the volcano with park officials, hiked around the rim, and was stunned by what she saw.

"I felt like I had stepped forward through time." Endemic scalesia trees sprouted up through once denuded earth. Shrubs that goats had stripped down to the bark only a month earlier were turning green and providing shade for sleeping tortoises. "Darwin's finches sang from the foliage and already a few shy ferns were appearing in damp hollows. For the first time in two decades I returned from Alcedo feeling like I was walking on air."

It's now three years since that glorious day, and I have organized a trek up the volcano with de Roy and two others to look around—and for me, finally, to visit the Galápagos tortoises in their habitat. Nothing in our wildest dreams could have prepared us for what we were about to encounter in the land of the giants.

CHAPTER 10

Walking with Giants

"We called him Tortoise because he taught us."

—Lewis Carroll

Tui de Roy knows Volcán Alcedo and its giant tortoises better than most people on Earth. The world-renowned wildlife photographer grew up in the Galápagos and has climbed this mountain on Isabela Island perhaps 80 times over the last four decades, up through prickly pear cactus and acacia, into forests of pungent palo santo trees and lichen-draped crotons. She has stared from the volcano's rim down into the crater, an alien tapestry of black basalt, white rhyolite, and crevices flanked by a tangle of green. She has camped on unstable ground near steaming fumaroles on the western side of the caldera and has spent weeks at rain-filled pools where the tortoises congregate. The volcano is visually deceptive. It rises to 3,650 feet but its shallow crater is only 270 feet deep.

Alcedo is home to the largest population of land tortoises in the Galápagos, the descendants of a single species, scientists say. They know this because the tortoises carry in their DNA genetic proof of an eruption that occurred on Alcedo about 100,000 years ago. These behemoths are the master lineage that survived the blast.

In May 2008 I travel by speedboat from Puerto Ayora to Isabela Island with de Roy, her partner, Alan, and a photographer named Pete

Oxford, who had come over from Quito. In all my visits to the islands, I had never seen tortoises in the wild, only in rock pens and on private land. I'm thrilled to be here with de Roy. She's one of my heroes, a photographer who captures the ethereal beauty of nature in all its perfection. She's a former naturalist guide here (one of the earliest), and she has written and photographed numerous books on the Galápagos, the Andes, Antarctica, and New Zealand, which she now calls home. Her goal on Alcedo, like mine, is to document how the volcano has changed in the last few years since teams of sharpshooters and specially trained dogs, working with the Galápagos National Park and the Charles Darwin Research Station, eradicated about 100,000 feral goats from Isabela Island. The wild ungulates had taken over and were competing for the same grasses and shrubs the endangered tortoises depend on for survival. In some areas, the volcano was so denuded it resembled a desert. Most of the shade trees were nibbled to the roots, and tortoises were dying of heat stroke or starvation.

Now, as our *panga* arrives at Shipton Cove, a narrow strip of beach on the eastern side of the island, the sun beats down, burning our faces. We're about one-half degree south of the Equator. De Roy is anxious to get going. She's a tall woman, with a boyish haircut and burnished skin that attests to a lifetime of working in the sun. She's wearing a khaki-colored shirt, gray shorts, and dark-blue sneakers. De Roy caches several gallons of water on the beach under a low-growing bush for our return trip. Then we adjust our heavy backpacks and follow the trail past tall prickly pear cactuses, some still in bloom. The stark terrain, the crystal air, the emptiness of the land remind me of the Sonoran Desert in southern Arizona, where I lived for nearly 30 years. As we ascend the volcano it becomes clear that Alcedo is to de Roy what the Sonoran Desert is to me: Home.

The trail climbs gently through the spare terrain of the coastal zone. The rains have been heavy this year, and normally dry palo santo trees are so leafed out I barely recognize them. Mockingbirds sing from their branches. It's getting hotter, and an hour into the hike we stop in some scant shade to rehydrate. I rest, but not de Roy. She's out in the sun with her Nikon, photographing a carpenter bee as it pollinates the waxy blossom of a prickly pear cactus. The fuzzy black bee is a stark contrast to the bright-yellow flower. Soon, a butterfly the color of a lemon enters the frame. De Roy waits for the decisive moment, then shoots. It's a close-up, perfectly timed. She steps back, visibly pleased. "I got my photograph," she reports, joining those of us still huddled in the shade.

Tui de Roy was born in Belgium in 1953, the daughter of what she calls "original free spirits." In the postwar atmosphere of Europe, her parents couldn't find the space and self-sufficiency they needed to live off the land. They'd heard about several European families who'd moved to the Galápagos and were farming, catching their own fish, and educating their children at home. So they moved to this little-known archipelago when de Roy was two years old. During the first year her parents carved a little place out of the wilderness up in the misty highlands on Santa Cruz Island, where they lived in a tent. But the perpetual rain and mist weighed on the family, and they moved down to the coast, where they built a tiny house and a fishing skiff.

Her father was a naturalist by vocation, and an artist. "He was very interested in wildlife and the ways of nature," she says. He was also hooked on photography and brought with him an ample supply of black-and-white film, reels, and chemicals. "He actually made black-and-white slides, processed the film in the sea. I remember he tied the film to a mangrove root and let the tide rinse the emulsion. There wasn't enough fresh water around to waste on processing."

By the age of 12 she was borrowing her father's camera, following his mentoring on what made a good photo—and what didn't. "He gave me a wee camera and I remember it had a little bellows and you had to cock the shutter and then click it and set the distance manually." But the budding wildlife photographer was more interested in shooting in color. She'd hit up visiting scientists or travelers passing through on yachts and ask them for film. "I loved climbing around in the mangroves or taking a rowboat around and looking for heron nests and pelican nests or yellow warbler nests, and investigating things."

In 1969 she was selling cured goat skins to the few tourists who came on the supply ship, saving up to buy a single-lens reflex camera. That year a filmmaker named Jack Couffer, who was working for Walt Disney, came to the islands to produce a piece on Darwin. De Roy took him up Volcán Alcedo. The rains were relentless, but something clicked: Couffer must have appreciated the teenager's knowledge of this stark terrain and her love of photography. He wanted to upgrade his still camera, so he gave her his old one.

"That was basically the beginning of my life, my career," she recalls. Then in 1972, Les Line, editor of *Audubon* magazine, visited the Galápagos. De Roy had recently been on Alcedo for ten days with a tortoise researcher and had taken some stunning photos. The captain of Les Line's boat had seen the photos. "He told Line, 'You must see these photos.' And he was like, 'Yeah, yeah, yeah. A 17-year-old girl and I'm gonna see her photos.'"

But the captain would not give in and arranged for de Roy to come aboard. She arrived in a dinghy, a young girl clutching her work to her chest, ready to present a slide show. "He [Line] just sat there in silence as I went through the photos. When we were done, he said, 'Can I take these with me? I'll let you know.'" The next year, in 1973, she got a phone call from *Audubon's* photo editor—and the promise of a plane ticket to New York City. Of course she went. Line and his colleagues were so impressed that they gave her a cover story—a photo essay of her images from Alcedo of giant land tortoises clustered in pools in the volcano's crater and a quality of light that seemed otherworldly. She had made it as a hot new wildlife photographer, and she wasn't even 20.

Now, as we climb the volcano that made de Roy famous, sweat drips from our foreheads into our eyes. Nettles stick to our socks, pricking our ankles. We use walking sticks made by a friend of mine in Puerto Ayora to whack down spiderwebs that cross the trail. Another hour later we again stop in some spotty shade. De Roy points out a Galápagos hawk that has just landed in a tree less than a stone's throw away. The raptor watches us for at least ten minutes. "He's just curious," she says. Sunlight glints off his hooked yellow beak, his brown and white feathers, his intelligent eyes. Soon a shrill whistle fills the air. "That's its mate," says de Roy.

Back on the hiking trail, we pass several sun-bleached skulls of the feral goats that once dominated this island. One skull has been placed in the branches of a tree, its empty eye sockets staring into infinity. The terrain changes yet again as we enter a grassy area—a stark contrast to the arid zone down at the coast. I stop to observe tiny black butterflies swarming a yellow-flowering bush. I've lagged behind the others, and just as I catch up de Roy, exclaims, "Greeting party!" There before us, munching on grass, is the first giant land tortoise we encounter on the volcano. It hisses and retracts its head as we pass.

We're entering a different realm, a moist upland habitat where the absence of feral goats and the presence of heavy rains have created a jungle. Where a dry trail once led up a relatively gentle slope, the vegetation is so thick it has swallowed the path. The ferns are so dense we can't get through. Fortunately, de Roy has brought a machete. She pulls the curved instrument from its leather sheath and begins whacking away. Blood trickles down her bare legs where cat's claw has snagged her skin, yet she never complains. She knows and respects this volcano and its wildness. It's like a blueprint engraved in her cells, yet on several occasions we lose the trail completely. Like the rest of us, de Roy soon becomes frustrated. She knows exactly where the trail *should be* and continues swinging the sharp-edged blade.

Soon we come to the first of three water caches set up by Galápagos National Park rangers. We're running low on water, but this cache won't do. Several finches float, dead, on filmy black slime. For three more hours we climb steep hills where lichens hang from trees all the way to the ground past vines tough enough to swing on. This is not the volcano de Roy remembers, and she's as shocked by the dense vegetation as the rest of us.

The second water cache is no better than the first. Invasive geckos have contaminated the tank. By now we're dead tired and thirsty. Even so, as we pound ahead, we discuss the many paradoxes in the Galápagos. I ask de Roy if she thinks the islands can survive. "It's already too late," she says without hesitating. "There are far too many people here." Nevertheless, she believes the park service has done a good job of keeping remote islands like this one relatively wild. "People could ride up here on ATVs (all-terrain vehicles), but it's not allowed. Can you imagine *that* kind of tourism in a place like this?"

As we continue bushwhacking up toward the volcano's rim, I begin wishing those feral goats were still here to serve as four-legged locusts. Then I catch myself. The fact is, Project Isabela was so successful that it's now almost impossible to ascend the mountain. The message: Maybe humans don't belong here.

It's getting dark and a thick blanket of mist rolls into what has become a jungle. We've entered yet another life zone where mushrooms grow beneath vines that trip us up, sending us face-first into the underbrush. Crotons, tall shrubs with leathery leaves, weep resin onto our clothes and stain them black. We must tread carefully. The trail is muddy and strewn with loose volcanic rocks. Soon the light vanishes completely. We've been hiking since morning, and now we're inching our way straight uphill in the dark of night behind de Roy's machete, trying to reach the rim and much-needed water. Finally, we give up. Our *machetera* should be exhausted, but she still has enough energy to help the rest of us bivouac in a tick-infested clearing under moss-draped trees. We have almost no water, but alone in my tent I console myself: Tomorrow I'll walk among giants.

Morning. I wake at five o'clock to the sound of someone yawning. It's de Roy, up in the croton bushes with her machete, clearing a trail. As she swings the blade, dew showers her mud-stained clothes. She must get to the rim and then hike a further 20 minutes or so, to the third and final water cache at a shack called a *caseta*. If the tank is empty or polluted we could be in trouble. But fate is kind: The tanks hold an abundance of cool, clear water and de Roy returns to camp with a gallon bottle for us to share.

The sun has not yet risen as we climb through mist thicker than potato soup. She plods ahead. When the rest of us catch up at the rim, she's perched on a boulder, smiling. Three young Galápagos hawks circle above us, as if in greeting, then disappear into the mist that rises from the volcano's floor. The moment lingers for a while, this lightness of being in contrast to the heavy weight on our backs.

Along the rim, more goat skulls appear, and tortoises wallow in brick-colored dirt. Most hiss as we pass. The larger tortoises have left scat the size of eggplants: undigested grass. Tiny white mushrooms grow on some of the droppings. The view down into the crater is still partly obscured, but from time to time, the mist rises and one of the island's greatest spectacles appears: lava flows and great fissures caused when the crater floor dropped; a forest of impenetrable green; and sulfurous gas spewing from heat vents—Alcedo's lungs. In some places we can see the tracks made where tortoises slipped to their deaths while traversing earth eroded by thousands of grazing goats.

De Roy sheathes the machete. Up here, the namesake creatures of these islands have blazed a trail. Oxford, who is also a former naturalist guide in the Galápagos, begins photographing a behemoth standing near the edge of the rim. Its face resembles that of the whimsical character. E.T. Oxford's timing is perfect: The sun emerges from the mist, casting light over the island all the way to the sea. He, too, is stunned at the sudden radical change in the landscape. "I've never seen so many tortoises up here on the rim," he says, shaking his head.

Oxford, who is British, lives in Quito with his wife and partner, Renée Bish. He's tall and muscular with wavy brown hair that's beginning to gray. He's dressed in camouflage pants and a T-shirt. And although the sun burns down, he wears neither a hat nor sunglasses. Most of his work focuses on indigenous cultures in relationship to the environment; so far he has published nine books. The cover of his photo book on Ecuador's Huaorani Indians (a traditional rain forest culture still battling oil exploitation in their homeland) shows a muscular hunter with a bowl-shaped haircut, naked except for a piece of string supporting his penis. The image is blurred, revealing action as he hurls a ten-foot-long spear while running barefoot through the jungle.

At the caseta we survey the descent into the crater. It looks deadly, another long hike with a bushwhack down a precipitous slope where fumaroles belch volcanic gas. My body aches, and the prospect of more trailblazing through a jungle infested with ticks mortifies me. I'm highly allergic to flesh-biting insects, and my once broken kneecap is screaming. The crater floor is a waterless hell, which means hauling down a lot of water to keep us hydrated for the next two days. I begin, silently, to balk.

We brew coffee on the gas range inside the caseta and sit around on plastic chairs. The wooden floor is a mass of dried mud and mice droppings. The coffee is hot and feels good in the cool morning mist. De Roy's partner, Alan, is an electrician and former beekeeper from New Zealand whom she met on the Internet. As we discuss our plans, he says, "No one, not one of us, expected these conditions. Here I am wearing $300 hiking boots and Tui's wearing $3 sneakers. She can outdo me anytime." The truth is that Alan seems a good match for de Roy. He has held his own. He helped slash through some of the toughest brush on this volcano in the dark of night while heaving a backpack that weighed half his body weight. It included my small day-pack. Oxford, who was straining under his own camera gear, carried my tent. Ashamed, I confess that I'm not tough enough to descend into that nebulous hell, not like de Roy. I gaze down at her swollen ankles. They're covered with tick bite upon tick bite, and her skin is perforated by nettles, yet she's ready to rally.

De Roy agrees that the volcano is radically different from anything she ever imagined. "It's just hideous," she says. "I'm afraid this may be my last trip up Alcedo." Under these conditions any normal person would give up and sleep on the panoramic rim, or return to the beach and camp in the sand where a cool tropical sea breeze blows off the eastern side of the island. But de Roy is no ordinary human being. As a wildlife photographer and naturalist, she prides herself on pushing the limits. I would see her endurance later, on Española Island, where we camped with a biologist to observe the endemic waved albatross.

My plan on Alcedo from the very beginning had been to follow de Roy down into the volcano and document her activities as she photographed giant tortoises in rain-filled pools. Instead, I embrace my limits and pitch my tent on the grassy rim as my companions enter a small forest of trees, each one carrying an extra gallon of water. Here, on the volcano's edge among knee-high ferns, I'll spend the next two days surrounded by tortoises, which amble by as though I'm invisible.

Volcán Alcedo is home to the healthiest population of giant tortoises in all the Galápagos. About 5,000 of the reptiles, about one-third of the total number in the islands, live here. These tortoises can weigh more than 500 pounds and live more than 150 years. Mating usually occurs in March or April, when the grunts of copulating males are so distinct that park rangers call this the loudest sound in the Galápagos. Once inseminated, the female ambles off to dig a nest in soft earth, where she can lay up to 20 eggs. Then she covers the clutch with dirt and urinates on it to cement it down. The hatchlings are born about three months later. Like sea turtles, they're extremely vulnerable; they must find cover before hawks, owls, and feral mammals such as cats, dogs, and rats nab them. Even the adults are prey—for poachers. Throughout the years, scientists have found dozens of slaughtered tortoises on Isabela Island, victims of an influx of fishermen and colonists who recently arrived in the Galápagos. About 70 percent of the human population is new to the islands. Most of these newcomers don't follow sustainable fishing practices in the protected marine reserve, authorities say, neither are they aware of the biological diversity that surrounds them nor how fragile it is.

Morning garúa fills the crater from rim to rim. It looks like snow, as though I could ski across it to the far western side. But by midday it lifts like a veil, exposing a lost world of smoking fumaroles where minerals have painted a palette of salmon, tan, and lime. Rivulets of green appear in volcanic crevices. All day long the mist creeps in and out of the crater like dragon's breath: the Earth as living entity. This, to me, is the real Galápagos, that illusory, hardscrabble world that empties into the mystic. As I peer into the crater through my binoculars I can see tortoises huddled in rain-filled pools, seeking reprieve from the heat and the blood-sucking ticks. This is where my hiking companions have pitched their tents. Their plan is to work in the pristine light of dawn and dusk in those rare moments when the sun breaks through. Here, in the bowels of the volcano, they can smell the sulfur from the nearby fumaroles. And all night long they listen to farting tortoises. As de Roy later recalls: "It was hilarious. They have a pretty rudimentary digestive system. They churn up these ponds to the consistency of chocolate sauce, but it doesn't smell like chocolate sauce. All of that fermentation causes plenty of gas, and so you hear this *blub-blub-blub* [and you see] these ripples coming out and creating rings around their backsides."

Up on the rim in my own camp, a different microcosm unfolds. In the morning I watch a saffron caterpillar, thin as a piece of spaghetti, wriggle up a blade of grass. As I follow the trail I come upon two tortoises mating. The male is much larger than the female. He has pinned her against a tree, his alien neck stretched over her shell, bellowing at her withdrawn head. Farther on, two tortoises greet each other face to face, raise their necks, and touch noses as if they're about to spar. But the larger one retreats into the shade while the smaller stays put, chomping grass as intensely as a goat. She doesn't pull her head in as I watch, but turns to face me, the sun glinting in her primeval eye.

Then something catches my own eye. It's a quadruped, a mammal, and as I turn to investigate, it slinks behind a rocky outcropping. It looks like a feral dog, one that escaped its master during the goat eradication project, but I don't get a clear enough look. Later, when I report this sighting to the Galápagos National Park, I learn from one of Project Isabela's directors that it was a feral goat—one that got away.

My third day on the mountain I hike along the rim to meet up with de Roy and the others. I'm carrying a gallon of water in case they've run out. To the east, tiny Rábida Island appears, and the much larger Santiago Island rises to the north. If I walked to the western edge of Alcedo, I would see Fernandina Island, the youngest and most active of all the Galápagos volcanoes. I'd be looking across Urbina Bay on the western shore of Isabela, an area that underwent volcanic uplifting in 1954, and where today, visitors can snorkel with sea turtles, marine iguanas, and rays.

About an hour into my walk, I meet the campers as they emerge from the trees. De Roy isn't even winded after the climb. Furthermore, she has plenty of extra water in her jug. She's astonished at how lush the volcano is now that the goats are gone. From time to time, she stops to marvel at club moss and shelf mushrooms. The endemic scalesia has returned, along with giant ferns. As I follow her she remarks on how healthy the tortoises are. "Now, when they pull their heads in, there's a rumple because their necks are so fat." She gazes down at a grazing tortoise right beside us. "That's right, my friend," she says: "Eat to your heart's content."

De Roy is triumphant: The volcano has been liberated. "The tortoises were enduring. They were hanging in there waiting for deliverance, and their deliverance has come." To emphasize their victory, she shows me where several tortoises slipped to their deaths after goats destabilized the ground. "The rocks had no vegetation to cling to, and the pebbles were as slick as ball bearings." Sure enough, at least three huge carapaces lie sun-bleached on the crater floor. As we move on, it becomes clear that she's in her element as she stops to photograph a close-up of a tortoise or a moth, seeing the story in the details. Her modus operandi in photographing wildlife: keeping still. "It's like Soto Zen," I suggest. Just sitting. De Roy replies, "The Zen concept is a good reference point. Photographing puts me in a state where everything is at peace, where here and now is what matters. All of my attention is focused on what is there in front of me."

"What is it that drives you?"

"Beauty, peace, harmony."

And focus, the perfection of nature captured by an untrammeled vision. "It's my mantra, if you like. Meditation is not easy, and a mantra helps. Photography is the reason I'm still there three hours later. It's the reason I don't mind being cold or whatever because when I look at my pictures and I see a photo that really came out, I have this buzz of pleasure seeing what I regard as the beauty of a reflection of harmony."

She's referring to that perfect second when the sun glints in the eye of a red-billed tropicbird, or a Galápagos hawk lands on the back of a giant tortoise, or a lava lizard nabs flies off a lazing sea lion with a lightning-quick tongue.

On the last day we hike back down through the brush past more goat skulls. De Roy is hoping to find a colony of huge land iguanas, but it's too hot and they're hiding out in the shade. Oxford and I continue down to the beach, dump our packs in the sand, and dive into the ocean. It's the first bath we've had since we arrived here four days ago. When de Roy appears with Alan, she discovers that her water cache on the beach has been stolen. She's not happy, and when our speedboat arrives to take us home, she asks the captain to pull up alongside a traditional fishing boat just off shore. The boat is old, its blue-green paint peeling off the hull. The crew has been fishing for bacalao, known as grouper in the Galápagos, and is drying their catch in salt. As we pull up, de Roy asks a man with a few missing teeth if the crew has taken our water. "No, no," he responds. "We haven't even been to shore." It's obvious he's lying; he won't even meet our eyes. But the fishermen of Isabela are quite another story.

CHAPTER 11

Pirates of Villamil

"The fishers also shall mourn, and all they that cast angle ... shall lament, and they that spread nets upon the waters shall languish."

—Book of Isaiah 19:8

Puerto Villamil is a sleepy fishing village on the far western island of Isabela. To most tourists, this town of 2,500 residents feels like the *real* Galápagos. The streets are made of sand: You can walk across town barefoot. Not long ago spiny lobsters were so plentiful a person could reach down into the turquoise surf and pluck one out for dinner. It's a town where palm trees rustle in the breeze, marine iguanas laze on rocks, and tiny penguins share the waters with sharks. At high tide fishermen play checkers with beer bottle caps, and everyone plays bingo with popcorn kernels.

This may seem like paradise, but a closer look reveals a town full of paradox. The sand for these idyllic roads (and most of the houses) is bulldozed from the beach, an activity that destroys the nests of marine iguanas and causes erosion. Lobster is now so scarce that individuals of the species are tagged and monitored with global positioning systems (GPS) from afar. The palm trees were introduced after the mayor visited Miami Beach and thought they would spiff up the town. With the palms came another bright idea: globe-shaped lamps attached to wooden posts that stay lit all night long, blotting out the stars and annoying beer-drinking tourists. The beer, by the way—and the popcorn—is barged in from mainland Ecuador on subsidized fuel.

The truth is that Villamil, despite its laid-back charm, is not as peaceful as it seems. The village has made international news over the years when fishermen rioted after officials refused to cave in to their unreasonable demands. The headlines read: "Turmoil in Paradise," "Unnatural Selection," and "Homo sapiens at War on Darwin's Peaceful Isles." Here's how it unfolded.

In November 2000 Juan Chavez, the local director of the Galápagos National Park, thought he was going to die. A throng of angry fishermen ransacked his house in Villamil and tossed his young daughters' toys and a baby's crib out into the street. Then they torched his home and all his belongings. Pandemonium broke loose throughout Villamil when the motley mobs moved on to set the Park Service Headquarters ablaze, after breaking windows and toilets, destroying computers and years of scientific data. Now Chavez was running for his life. Perhaps it was pure adrenaline, but somehow he managed to escape and was hiding out in a mangrove swamp at the edge of town. Assured that his wife, Martha, and their children were safe, he would stay there among the poison apple trees, ghost crabs, and marine iguanas that slithered through the water like snakes until a Navy boat arrived offshore to rescue him.

The revolt began when more than 900 fishermen grew angry over limits the Galápagos National Park had placed on lobster quotas. They destroyed some national park offices on Isabela and took over others. They torched a park vehicle and blockaded roads. They also took ten endangered tortoises hostage from the park's Tortoise Breeding Center in Villamil, and they destroyed incubators that contained tortoise eggs. Park employees went into hiding while a small unit of police tried to keep the fishermen at bay. The Ecuadorian government ordered immediate release of the tortoises and the departure of all fishermen from the park offices. The fishermen complied, and then it was "business as usual": The fishing quotas remained the same and the ringleaders got exactly what they demanded, while environmental laws intended to protect the marine reserve were ignored by government officals employed to enforce them.

In a radio interview soon afterward, park naturalist and wildlife photographer Tui de Roy warned that overfishing can cause a collapse in the marine ecosystem. "A more significant problem is the fact that the fishermen are proving time and again through their actions, that have been going unpunished, is that they can take the law into their own hands. … We're talking not just about quotas, not just about the sustainability of lobsters or sea cucumbers or even sharks, we're talking about … total disregard for any kind of laws and regulations to protect the island."

Villamil has long been a hotbed of unrest among the archipelago's fishermen, a sector that considers the park service and conservationists "the enemy." They call themselves "pirates," outlaws determined to get what they see as theirs. Backed by powerful political interests in both Galápagos Province and on the Ecuadorian mainland, they succeed. "At the moment there is absolutely nothing left at the park office," park director Chavez told journalists after the Navy rescued him. " 'We don't even have a pen to work with."

Within in a week a group called the Committee of Concerned Galápagos Citizens issued an international press release that began: "It is inconceivable that at the beginning of a new millennium, in such a special place as the Galápagos World Heritage Site, the terrorist actions that have occurred since Monday can still happen. … How can a small group of illegal outsiders, posing as Galápagos fishermen, commit all kinds of criminal acts for years, without any retribution?"

A few days later an independent union, the Galápagos National Park Wardens Association, sent an ultimatum to the President of Ecuador demanding an immediate end to the violence and warning that, without the government's intervention, they would be unable to protect the park or the conservation laws of Ecuador. Throughout the week civic leaders had received anonymous death threats, and Eliécer Cruz, director of the Galápagos National Park, continued to plead for the arrest of the ring-leaders on Isabela Island. Ironically, his main opposition came from the governor of Galápagos Province—the archipelago's highest official— and a few of the islands' mayors, who had helped lead the uprisings. In essence these officials, in condoning violence, provided the impetus for *el piratismo* (pirating) in internationally protected waters.

This wasn't the first time machete-wielding fishermen had revolted when their demands weren't met. In 1995 fishermen rose up twice over a ban on fishing for a species of sea cucumber, *Isotichopus fuscus*, which is sold on the Asian black market as an aphrodisiac. (The sea cucumber trade once brought in more money than could be earned from dealing cocaine.) The *New York Times* reported that in September 1995, "radio broadcasts [on the islands] urged looting and arson, while ordinary people vowed to become guerilla fighters." In 1996 the fishermen again hit the news after slitting the throats of several endangered tortoises on Santa Cruz Island. In 1997 they revolted when the quotas returned. Then in 1999 they stormed the airport on Baltra Island and grounded the environment minister's plane on the tarmac. They shot at research-ers and park guards, taking some of them hostage. Once again they seized endangered tortoises and threatened to release goats, cats, dogs, birds, and invasive seeds on the islands if their demands weren't met. Quotas were immediately dropped, few people were ever arrested, and the federal government never stepped up security.

The Galápagos Marine Reserve (GMR) is the second largest in the world at 50,000 square miles, an area about the size of New York

State. (Australia's Great Barrier Reef Reserve is the largest.) Established in March 1998 by the Ecuador government, its sheer size makes the reserve nearly impossible to monitor. Until recently, the park had a single boat to patrol the entire area, which is one reason why fishermen continue to ransack its living treasures, nearly depleting sea cucumbers, killing sea lions for their penises or cutting them up as bait to catch sharks, and using illegal longlines to catch those sharks for their fins, teeth, oil, and skin. By-catch caught on the huge hooks attached to these lines often includes the endangered sea turtle and the waved albatross, which can see the bait from the air. Later I would see proof of this danger on Española Island, where I met a biologist who found an enormous hook lodged in an albatross's throat. The bird was still alive.

In the summer of 2007 I visit Juan Chavez, the Isabela Island park director who hid out in the mangroves in 2000. I've come to Villamil with my colleague Diego Quiroga from the Galápagos Academic Institute for the Arts and Sciences (GAIAS), who wants to catch up on the news. Chavez is thin and athletic, a quiet man with slicked-back hair and wire-rimmed glasses. The blinds in his tidy office are drawn, and although it's wickedly hot, the air-conditioning unit is turned off. A poster of a sea lion and a shark covers one wall, and above the desk is a wall calendar bearing a picture of Jesus. We ask him about the riot in 2000, but Chavez is a modest man and doesn't say much.

"The most important thing is to be calm, to have a good attitude. Even the volume of your voice is important when you're talking to a large group of angry people," he offers, and then falls silent.

When I mention the photograph in a Quito newspaper that shows his daughter's crib smashed in the road, he admits, "We lost everything."

"Do you know who did it?"

"I don't know who it was because I never investigated. If you're just thinking of material things, your life will be so sad. I told my wife, 'Please, my love, think. There are people so poor. There are worse things in the world.' "

A year earlier, while I was teaching for GAIAS, I had met Chavez's wife, Martha Véliz, who was running the only school for environmental education in Villamil. At the time I was unaware she was Juan Chavez's wife. Here she was, teaching the children of the same fishermen who'd burned down her house. Here was courage. Here was extraordinary dedication.

Véliz is a stunning woman whose wavy black hair hangs over a light-green T-shirt. As the director of the Charles Darwin Foundation's (CDF) Centro de Educación Ambiental (Environmental Education Center), she helps teach the 90 school-age children of Isabela Island why conservation is so important here. We sit at a table beneath colorful mobiles of the islands' wildlife, all of them made by her students. "The vision," she says, "is to educate people about biological diversity [in conjunction] with social and educational activities, and to simplify the science in a language that people can comprehend." Children learn in three stages: The youngest, known as *las iguanitas* (little iguanas), learn the basics, like how the islands were formed and how species first arrived here. The middle-school children, *los piqueritos* (little boobies), learn more about science and natural history. The older students, *las tortugas* (tortoises or turtles), go into the field every week to study sea turtles or giant land tortoises or the biology of marine iguanas or the mechanics of volcanism. When Sierra Negra Volcano erupted in October 2005, Véliz took a small group of students up the volcano's slope to watch from a safe distance.

"What did they think?" I ask.

"Wow!" she says, her dark eyes flashing. "WOW!"

"Do any of your students want to be fishermen?"

"No, none of them," she says emphatically. "Even their parents don't want them to be fishermen. They don't think there's a future in fisheries." The irony is that the program has been so successful that many fishermen—those crazy pirates—now bring their children to this school. Some parents even participate in the field workshops now that they've out-fished their livelihoods and turned to something they know little about: tourism.

A few months after I visited Véliz at her school, park wardens found the remains of four butchered land tortoises. Poachers south of Villamil had harvested two adult females, an adult male, and a young adult. The news came on the eve of an island tour by the Friends of the Tortoises, a children's group from Véliz's environmental school. Véliz put a positive spin on the massacre: It could have been worse, she told the CDF. "The kids' efforts in creating awareness about the importance of the giant tortoise [in] their home community has helped reduce the killing from almost 50 in 2003 to only 6 this year."

At the time, a pilot project on Isabela Island placed young tortoises in the care of local farmers as a way to educate them, protect the species, and provide the farmers income through ecotourism. The students from Friends of the Tortoises were helping out as volunteers at the captive-breeding center and as guides at the farms to teach locals that protecting the tortoises is more important than eating them.

In December 2007 I travel to Villamil to spend New Year's Eve and check into a new hotel on the beach, the Hotel Albemarle, owned by a Briton named Max Murray and his Ecuadorian wife, Diana. All across town residents are preparing for New Year's Eve, a raucous celebration that resembles something out of a Gabriel García Márquez novel. On every sandy street people have constructed effigies of friends, neighbors, foes, cartoon characters, and political upstarts. The mannequins are made from old clothes and are stuffed with sawdust, newspaper, and firecrackers; their heads, some quite artistic, are made of papier-mâché. Townspeople scribble silly jokes on posters; poke fun at friends; write prayers of forgiveness and place them next to the effigies; and then burn the whole mess at midnight in a grand explosion of flames.

Outside the police station an effigy of a cop in a blue uniform is propped up against a lamppost. Down the street a bunch of overstuffed drunks slump over a table filled with empty beer bottles. Other characters aren't so comical. At midnight effigies of George W. Bush, Joseph Stalin, and Ecuador's President Rafael Correa will all go up in smoke. It's social and political street theater at its best.

Across from the town square I stop to observe an egg-tossing contest in which people pair up, stand about 12 feet apart, and hand toss an egg back and forth. The idea is to avoid breaking it. The announcer blares commands through a microphone. "*Listos?*—Ready?" "OK, hold your *huevos*," he says, using the slang word for testicles. "Now toss your huevos!" The crowd roars. Two women are in the lead. They handle that egg like it were a newborn babe. First, it lifts off the hand of one woman as gently as a butterfly. Her partner, perhaps her daughter, locks eyes with the object and catches it intact. They're on a roll now, and the announcer coaxes them to continue, but now they're the only contestants left, so they stop to receive their prize. It's something made of plastic, though I can't make it out through the throng of onlookers. Meanwhile, a boom box at a nearby bar blares out Pink Floyd: "We don't need no education. We don't need no thought control."

At sunset the rest of the town comes out to play. Parents promenade with children dressed in their Sunday best. Then there are the *viudas*—widows—men who dress in drag, all in black. It's a long tradition in

Ecuador. I pass several of them as they stop and joke with their friends. One has "breasts" made of balloons and something stuffed up his miniskirt to enhance his backside. Black curls cascade down his back as he struts around in stiletto heels, carrying a stylish handbag. Another looks like an Amazon. His legs are the size of sequoias, hard and strong, his micro-miniskirt so scant you can almost see his crotch. He's beautiful in a kitschy kind of way, almost androgynous; his makeup is impeccable. But beneath this guy's guise is a local policeman who speaks to his uniformed buddies in an effeminate voice as they stand right next to the mannequin they'll torch at midnight. Welcome to the lighter side of town.

Around midnight, as the effigies go up in smoke, I climb to the rooftop of my hotel. From up here the village looks like a battlefield. Fireworks light up the town and the lava fields beyond, and cheap rockets whiz by like comets. The breeze carries the stench of gunpowder and beer. Tourists dance on the beach at Bar de Beto while couples steal off to make love in the shadows.

New Year's Day 2008. I wake early and walk the beach alone, as whimbrels peck in the sand, bright-red ghost crabs emerge from their holes, and marine iguanas sun themselves on rocks. Suddenly, a Galápagos hawk swoops down and snatches a baby marine iguana in its talons and sails west over the hung-over town. Villamil is dead silent. A few drunken fishermen stumble around, some still drinking, others falling off their bikes. A few have passed out at plastic picnic tables, or in the street, or in their hammocks. Stray dogs rummage through the trash-strewn town. Where the colorful effigies once stood, black soot has taken their place.

Six weeks later I return to Villamil to meet with some naturalist guides and dive masters who are also musicians. I also plan to interview the mayor of Villamil, Pablo Gordillo Gil, who was indicted for cutting down protected mangrove trees. Gordillo's father—Jacinto Gordillo Gordillo—is one of the most revered naturalists in Galápagos history, a former priest who lives in the highlands, where at age 84 he still cultivates endemic plants that are on the edge of extinction.

It's Valentine's Day and at Bar de Beto three of the most famous nature guides of the Galápagos, including Mathias Espinosa from Scuba Iguana, are playing a gig. I sit at a wooden table in the sand with my hotel-owner friend Max Murray and his sister and her boyfriend, who are visiting from England. The band tunes their instruments then dedicates the evening to all those *parejos* (couples) out there. Espinosa, who has cut several CDs with his biologist wife, María Augusta, plays guitar. A big guy, the diver Jimmy Iglesias, bangs on his bongos with a distinctive Cuban beat. A guide named Minino Bolona also plays guitar. All of them sing, but Espinosa is the lead.

After two Cuba libres I kick off my sandals and join some people dancing in the sand. Yes, this is the real Galápagos—a little piece of paradise lit by candles where the drinks keep coming and the ocean is warm enough to swim in. I feel a lightness return as I scuff down the beach under a sliver of moon, hoisting up my skirt and jumping through waves. I think of Puerto Ayora, an assembly line of tourists, cargo boats, and taxis, and think I'd rather live here. Then I catch myself. I'm just a visitor, a journalist passing through trying to understand the complexities of these islands.

When I return to the bar, the musicians are winding down for the night. Bolona plays a solo and sings it in English: John Lennon's "Imagine," the perfect ending on Día de Santo Valentino. Tomorrow I will talk with the venerable old priest up in the highlands. I've asked Espinosa to come along with me. He's spent years working as a professional journalist and still produces documentary films on issues about the Galápagos.

In the morning we hire a car and head up to an area near Cerro Verde. The sandy road is rutted from the recent rains and our driver must steer around enormous potholes. As the road climbs above Villamil the land becomes more fertile. We pass ferns, coffee plants, and trees heavy with grapefruits, guavas, and papayas. The car enters Rancho Negro, Jacinto Gordillo's ranch, and stops at a modest house with a corrugated tin roof. There to greet us is one of the last true pioneers of the Galápagos, a lifelong botanist who still works for the Charles Darwin Research Station (CDRS). His job is to propagate and distribute endemic plants, especially scalesia, which is threatened by invasive species and ranching.

Gordillo is a slight man with sunken cheeks and kind brown eyes that shine from beneath bushy white eyebrows. He's dressed in gray shorts and a yellow polo shirt that advertises: Galápagos Coffee. Gordillo's arms are as dark and wrinkled as the old land tortoises he once studied for the CDRS—a testament to a lifetime of working under the sun. He's still quite agile, but he's beginning to lose his memory. That's one reason he's writing a new book on the human history of the Galápagos.

"*Bienvenidos*," he says softly and invites us inside.

Jacinto Gordillo Gordillo was born on April 5, 1925. He came to Isabela in 1952 as a Franciscan priest, the first ever on the island, where he lovingly served the people for 13 years. Then, in 1964, a different emotion tugged at his heart. Gordillo left the priesthood, married, and eventually had five children with Gloria Gil, the great-granddaughter of Villamil's cattle baron founder, Antonio Gil, who lived here in the late 1800s.

Leaving the priesthood carried a stigma, however, and Gordillo had trouble finding work. Then, in 1966, when the CDRS set up an office in Villamil, he was hired to monitor giant land tortoises around Volcán Sierra Negra and Cerro Azul. That year he bought two hectares of farmland for 200 sucres. At that time there were no regulations; colonists could settle wherever they pleased. Today Gordillo owns 54 hectares, which he's split among his five children, including his son the mayor.

His house is eclectic but orderly. The living room contains a washing machine, a freezer, and an electric piano. There's a comfortable sofa with four matching chairs, and a dining room table where tall red candles rise from a candelabrum. A picture of the Last Supper hangs on the wall near a wooden hutch that supports a statue of the Virgin Mary, a rosary, and a gas lantern. We sit on the sofa before a coffee table covered with the books and articles Gordillo has written over the years and has hauled out to show me. There's a book on flamingos and wetlands and one on Galápagos botany. I'm glad Espinosa is here to help translate. As a naturalist, he knows the scientific names of plants and *that* can get complicated.

I tell Don Gordillo I've just spent time in a sea turtle camp at Quinta Playa. He knows the place well. "Are there many flamingos?" he asks.

"About 30," I reply.

"No!" he exclaims. "There used to be 300!" He's visibly upset.

Suddenly, a hard rain descends, pelting his tin roof like a barrel of BBs. Gordillo jumps up and runs outside to grab his laundry from the clothesline. The rain, fickle this time of year, stops almost as quickly as it began. I want to see Gordillo's experimental garden so he takes us outside. First he shows us a special garden in the front yard that contains both native and non-native plants. There's *chilca*, a flowering shrub from the sunflower family; at least three species of thin-leafed daisies; and a plant he can't identify so he is analyzing its DNA. Introduced croton grows here next to a plant with leaves the size of an umbrella. I know this plant.

"This one is *oreja de león* (lion's ear)," he says.

"Do you mean *oreja de elefante* (elephant's ear)?" I ask.

"Oh, that's right," he says, apologizing. "Oreja de elefante."

It's clear that Gordillo's memory, that fragile network of nerves, synapses, and sensations, is waning. I want to hug him, to reassure him that memory is both a blessing and a curse.

We pass through a gate into his organic garden. Compost piles appear everywhere. Gordillo stops beneath a towering grapefruit tree to point out Galápagos cotton, an endemic species whose bright-yellow flowers resemble hibiscus and whose seeds really do produce cotton. Here also is endemic *cafetillo,* white wild coffee, so named because it resembles coffee plants, and an endangered plant called *arbolito rojo* (little red tree). Cattle eat it, farmers bulldoze it, and Gordillo is nurturing it back into existence.

The old botanist bends down to pull up some weeds and tosses them to the side. This garden is an arboretum of native and endemic plants. He points out *rodilla de caballo* (knee of the horse), also known as vervain or glorybower; several species of *margaritas* (daisies); and endemic acacia and prickly pear cactus. But the most interesting endemic is the scalesia, the tiny daisy from the sunflower family that morphed into a tree. Like Darwin's finches, scalesia is a perfect example of adaptive radiation for its ability to survive in a variety of habitats, and in the process, evolve into a brand new entity. The genus here contains 15 species, 19 if subspecies and varieties are included.

To Gordillo, this mutated daisy is like a family member. Literally. In 1986 the naturalist was working in the arid lowlands of San Cristóbal Island for the CDRS. One day he discovered a new species. The plant, which looks nothing like a daisy, has blade-shaped leaves and clusters of tiny white flowers. It was named in his honor: *Scalesia gordilloi.*

As we pass through the garden I realize that Gordillo isn't wearing shoes, even though invasive fire ants march across the path.

"Why do you garden barefoot?" I ask.

"I like it," he responds. "It's healthy."

Back inside I ask Gordillo how things have changed since he came here in 1952. "The changes are huge. It was like a big family," he recalls. "We traded for fish and vegetables. Money wasn't important. Before, everyone had something to eat. Now at the dock you have to pay for fish. No money, no fish. Some people are too poor to buy it."

Gordillo blames part of this on vying political parties that are destroying the Galápagos. As for his son's indictment for cutting down mangrove trees, he claims that another political party is attacking Pablo because the opposition is envious of his success. "People are saying terrible things. He only cut branches, limbs."

Then, faithful father that he is, Gordillo lists the good deeds his son has accomplished during his seven years in office, although some aren't as good as they sound, such as potable water, the sewage system, and garbage collection. One visit to the Villamil dump is an absolute nightmare. Nevertheless, Gordillo ticks off the following list: His son, he says, reconstructed the municipal dock; created a viewpoint of the wetlands; built a wooden walkway; installed city lights; improved garbage collection; bought a bus to transport farmers up and down from the highlands; improved the sewage system; installed a potable water system; helped buy computers for 20 families; built a school in honor of his father (which does bear Jacinto Gordillo's name); and constructed an airport.

"What are your own dreams for the Galápagos?" I ask.

"It's hard to define that, but I'd like to return to a time when life was simpler, when there were few people here and it was peaceful. I dream about this all the time."

"Literally?"

"Yes, I dream about this more and more lately."

Gordillo walks into his kitchen and pours sweet white wine into goblets for Espinosa and me, then sits in a rocking chair smiling.

I ask this priest-turned-botanist if he'll join us.

Gordillo taps his temple and says: "*Mi memória.*"

I feel as though I've just been sanctified. When we leave I surprise myself by saying, "*Grácias, Padre.*"

Back in Villamil the next morning Espinosa and I walk down to the mayor's office. Pablo Gordillo is a robust man whose round belly pokes out over his jeans. His office walls look like a shrine to his accomplishments. Awards and plaques cover them from top to bottom. Espinosa and I sit on a yellow couch, against yellow walls, next to an artificial plant. Before he was mayor, Gordillo was a fisherman who shipped frozen grouper fillets to Proinsular, the main supermarket in Puerto Ayora.

The mayor says his major achievement was to pressure the government to issue new boat licenses. "Before 1998 there was a lot of conflict getting permits so fishermen could change their livelihoods and move into tourism." Fishing still occurs, he adds, but it's in crisis. During the sea cucumber bonanza Villamil had about 200 fishing boats. Today that number has been reduced to about 30. About 80 percent of the fishermen have changed to tourism without proper training or language skills. "The demand for transportation has changed because tourism has grown so rapidly," Gordillo explains. But the park is holding back on issuing any new permits, and that makes him angry.

"What's your relationship with the park?"

"Very bad," he replies, his dark eyes burning. The park and all those nonprofit foundations, he says, are more interested in "floating hotel" tourism than supporting the local population and teaching them about conservation. "The park, up to now, has failed, not just to protect an area, but to involve the community so we can grow economically and have tourism that includes the people."

Even so, his solution, he says, is ecotourism, which will soon become the town's principal industry and will open up pristine new areas around

Villamil. It's an implausible dream because Villamil cannot handle more tourism. The infrastructure isn't there, nor is a sustainable model for development. More tourism means that more outsiders will be needed in the service industry, and the government has cracked down on immigration. When the fledgling airport at the edge of town is expanded it could mean one more sacrifice zone in paradise by allowing in so many tourists.

Mayor Gordillo looks down at his watch. He needs to take his father to the airport.

"What really happened with the mangroves?" I chime in.

"This was a moment when everybody knew that UNESCO was going to declare the Galápagos endangered so the park was looking for guilty people," he claims. He then adds that the area in question was degraded with a run-down bar, litter, and cow pies. I rephrase the question:

"With all due respect, Mr. Mayor, what exactly did you cut down?"

"Do you want to see what was actually cut?" he snaps. "Let's go there right now and I'll show you!"

Mayor Gordillo leaps from his chair and storms out the door. Espinosa and I exchange glances, then follow him to his official staff car and climb in the backseat. Gordillo drives down a sandy road and stops near the main dock. He gets out and slams the door, his blue and white sneakers pounding the sand as he walks.

"I did not cut down any mangroves. They were just branches, and now they're recovering."

He goes on to tell us that they were infested with saltbush. (Saltbush does not infest mangroves, Espinosa later tells me.) The mayor says he's clearing the area to build a small ticket office for speedboats to Santa Cruz, a center for tourist information, and a store to rent snorkeling gear. It's all about ecotourism.

"Where are the mangroves that were cut down?" I ask.

"The Environmental Police confiscated the branches for evidence."

I don't tell him that I had watched a video of him and his workers cutting down trees, not branches, in Park Director Juan Chavez's office. Nor do I admit that I know who shot the footage and made it public: Henry Segovia, one of the fishermen who led the uprising here in 2000 while he was head of the town's fishing cooperative. It's a small-town thing. Segovia and the mayor are longtime rivals and share a vendetta like something out of an old Western movie. Apparently, Segovia owned a bar near the dock, but the mayor refused to renew his permit and asked him to leave. Segovia refused; he needed money to buy new engines for his speedboat. So the mayor threatened to bring in a bulldozer and flatten the bar.

"Henry's the type of guy that if someone is going to screw him he's not giving up," said a local resident who did not want to be named.

Revenge came with Segovia's mangrove video, a two-minute piece that screams with chainsaws and catches the mayor and his workers red-handed as the mangroves go down. Then the camera pans to a truckload of enormous logs. Segovia summons local park officials, who arrive on the scene. They demand that the mayor stop and tell him he's committing an environmental crime. "I could not give two shits," the mayor barks from behind dark sunglasses. "You're going to have to throw me in jail if you don't like it." Mayor Gordillo points his finger at Segovia's lens and snaps, "Turn off that camera!"

"I'm allowed to film this, *señor.*"

"You're not a journalist!"

"But this is a public place," Segovia retorts.

The camera suddenly goes cockeyed as though Segovia has been shoved, then the video goes black. Segovia immediately released copies of the footage to the media, and to Sea Shepherd Galápagos, the local chapter of the Sea Shepherd Conservation Society, a radical environmental group, which launched a campaign called Operation Mangrove. A district attorney from the Ecuadorian mainland flew in to the crime scene with then-head of Sea Shepherd in the Galápagos, Sean O'Hearn, and a team of forensic experts. They found that the mayor had indeed cut down white mangrove trees that were 80 or 90 years old and that he'd demolished a 1,000-year-old wetland by filling it with sand and volcanic debris. Wetlands are vital to the ecosystem and its endemic species. In destroying the wetland and its mangroves, the mayor had trashed one of the very attractions that ecotourists come to see.

Mayor Gordillo was arrested and indicted, but he fled Isabela Island to hide out on the mainland, where he rallied political support from his allies in the Social Christian Party before the indictment could be finalized. It was the first time a high ranking political official had been charged with an environmental crime in the Galápagos.

———————

Among the fishermen with boat permits, some have illegally sold or rented those highly coveted prizes to others for thousands of dollars a month. Others now haul tourists in unsafe speedboats from one island to another. A round-trip ticket from Villamil to Puerto Ayora costs $60. The maximum number of passengers is 16, but the total often exceeds that limit.

On my trip from Villamil back to Puerto Ayora on one such dangerously overloaded boat, a man who appears to be its owner gives some orders to the "captain" and crawls into the bow, where our luggage is stowed. There's so much stuff it covers the toilet for the two-hour journey. This is not good. The speedboat slaps the waves so hard that I sense my bladder could end up in my thorax. Several passengers hang their heads over the gunwale and vomit. The exhaust fumes from the outboard motors don't help.

At Puerto Ayora the boat's owner stumbles out of his little shelter, yawning and squinting in the morning light. He shows the captain where to dock and rubs his eyes. I recognize his voice from the mangrove video and his face as the fisherman-turned-tour-operator who in 2006 took my students and me to Las Tintoreras to see the whitetip reef sharks just off Villamil. It was the way he gripped the wheel of his boat that day, the way he maneuvered it out of the dock at low tide, the way he laughed (it was a rasp, really). But his eyes were a dead giveaway, revealing something of a rogue. Here was a guy, I thought, who probably lives on the edge, not out of necessity, but because he thrives on it. Here before me was one of the ringleaders of the 2000 Villamil revolt who sent Director Chavez running for his life, the same man who filmed the mayor cutting down the mangroves.

"I know you," I say, as he climbs over our luggage. "You took my students and me to see the whitetip reef sharks in 2006."

"*Sí*. You look familiar," he replies.

"You're Henry Segovia, one of the 'Pirates of Villamil,' right?"

I fear I've offended him, but he laughs out loud and gives me a hug.

Segovia agrees to an interview the next time he's in Puerto Ayora. When I greet him at the dock a few days later he's right on time. He's dressed in a Hawaiian shirt painted with bright red flowers and loose-fitting slacks. The scent of aftershave hovers around him like sugary fog. I look down at his cargo: boxes of Avon products. We walk down the street to a popular establishment called the Café Hernán. As we pass the Catholic church Segovia greets a priest standing in the doorway in a starched white robe that flairs out like a tepee. "*Buenos días, Padre,*" he says. The priest nods and returns the greeting. At the café I drink cappuccino made with coffee grown in the Galápagos. Segovia drinks mineral water.

I ask him what it's like being a fisherman these days, whether it's much more difficult.

"Yes, in reality it's very difficult because regrettably we've eliminated certain species. But there's an alternative that exists for the pescadores

of Galápagos: Tourism. What's standing in the way of that is the Galápagos National Park."

Segovia says that back in the early 1990s fishermen survived by capturing lobster. "We had a good life. The sea cucumber industry didn't
exist back then. We knew nothing about it." But in 1992 middlemen
started showing up from Asia, loaning fishermen money outright to
dive for sea cucumbers. Many became wealthy as they delivered the
illegal goods. Those who failed to deliver or pay back the advance might
never be seen again. But as the pescadores depleted the resources that
had lifted them out of poverty, the park cracked down.

"The park always says, 'Don't fish, don't fish, don't fish.' But they've
never given us alternatives. It's absurd, the system that exists right now."

Segovia takes a sip of water and stares out the window.

"Look," he says. "It's the park's fault and the fishermen's fault. Why
the park? For not permitting us to fish productively. How does one
make money if they keep restricting us? We have wives and children.
We're hungry. We decided that if they kept it up we'd have to fish illegally. That's why we fishermen are also culpable."

I ask if he considers the park the enemy.

"That's a good question. We've had so many problems with the park
service and their stupid laws. The laws here serve only certain people."

"Who?"

"The big tour operators. This has damaged the whole system and
left the fishermen out. It's very political."

And what about the politics of the 2000 uprising he helped organize
and lead in Villamil after the park placed quotas on lobster fishing?

"Did you burn down Juan Chavez's house?"

"Would a fisherman break children's toys and throw them in the
road? I wouldn't, but fishermen did. We destroyed computers. Yes, we
did that. And we burned a car and the offices of the park service. Why?
Because the laws are unjust—and also to engage [the park] in dialogue.

We had tried in a thousand ways, even with lawyers in meetings that began at three in the morning and lasted until three or four in the afternoon, hour by hour. And for what? So they could tell us, *no?*"

Segovia is staring at me, lost in that frenzied memory when he'd been pushed too far. "You want war?" he asks, "Let's go to war. *Revolución!*"

What strikes me about Segovia is both his candor and the way he dances around questions. In 1997 he was detained for fishing illegally for sea cucumbers and for carrying a firearm—crimes he emphatically denies. A car chase ensued in Villamil, and the police subdued him. They threw him in the back of the vehicle and removed his shoes so he couldn't escape over the hot volcanic rocks. But the ever sly Segovia found an opening. The first nanosecond the authorities looked away he flung open the car door and bolted over razor-sharp lava just like Godzilla.

Segovia had triumphed. He became the town hero because he'd defeated "the enemy."

CHAPTER 12

Neptune's Warriors

*"Who hears the fishes when they cry? It will not be forgotten
by some memory that we were contemporaries."*

—Henry David Thoreau

Farley Mowat is a devil of a ship, a black behemoth painted with the banners of every nation whose whaling vessels it has rammed or sunk. The 50-year-old ship is a former U.S. Coast Guard cutter with state-of-the-art communications, a helicopter, a colorful logbook, and a chart room that could be a museum. I got to board this flagship of the radical environmental group Sea Shepherd Conservation Society when it docked in the Galápagos in June 2007, just after UNESCO declared the islands endangered. About a year later Canadian authorities seized the *Farley Mowat* when its crew arrived to document fishermen bludgeoning seals with clubs and, sometimes, skinning them alive. The Humane Society of the United States condemns this annual hunt as the "largest slaughter of marine mammals on Earth." "During the course of this year's [2008's] commercial seal hunt in Canada, we estimate that sealers will slaughter 275,000 seals," the society reported. "This estimate is based on the 2008 harp seal quota."

It's not hard to understand Sea Shepherd's motivation. In my many visits to the Galápagos as a journalist and teacher I'd met the directors of the group and had spoken with the president and founder, Capt. Paul Watson, to learn more about their campaigns. Watson is a portly man with thick white hair, a white beard, and a hearty laugh. But beneath that Santa Claus veneer is a true warrior. Watson has no problem delivering the gift of, say, a ramming rod called a "can opener" or a stink bomb or a shotgun blast to a buoy whose longline hooks ensnare sharks, sea turtles, albatross, and other endangered species. Once the sailors from a whaling ship tried to board the *Farley Mowat*. Watson and crew aimed its cannon at the intruders and blasted them with 45-gallon shots of banana pudding. Sweet surrender, but these guys aren't fooling around. Sea Shepherd claims to have sunk ten whaling ships since 1979, including two in South Africa, three in Norway, and two in Iceland, which effectively put the Icelandic whaling industry out of business. In 1981 Sea Shepherd exposed illegal whaling in Soviet Siberia. And in 1991 the crew staged a different kind of protest: They boarded a replica of the *Santa María* during the anniversary of Columbus' so-called discovery of the New World to denounce 500 years of injustice to indigenous peoples. The Spanish government eventually apologized.

To say that Sea Shepherd's men and women are courageous is a gross understatement. The crew has braved 40-foot swells and 60 mile-an-hour gales. It has maneuvered around fluorescent-blue icebergs a hundred feet tall in temperatures that can kill. They've been chased by ax-wielding fishermen and have stood as quiet as mice as Canadian policemen held assault weapons to their heads. Some have spent time in jail. But not once has Sea Shepherd taken a human life, nor has the crew lost a life. The main philosophy of these "eco-pirates," as they call themselves, is to uphold and enforce international maritime law and to halt the illegal slaughter of wildlife. They even fly the Jolly Roger just for the romance of it. After all, says Watson, the crew isn't out robbing or looting; it's destroying property used in criminal activities and *that*, by definition, is not piracy.

Even so, the FBI has accused Sea Shepherd of "ecoterrorism," and the president of the Japanese Whaling Association lambasted the group for being "circus performers" and "dangerous vegans." The irony is that Sea Shepherd has been so successful in its campaigns that in 2005 this nonprofit group entered into a legal partnership with the Galápagos National Park. Strange bedfellows, you might say, but with common goals: to patrol the marine reserve and protect it from poaching, long-lining, and shark finning (the practice of catching sharks and slashing off their fins, then dumping the still living animals overboard to suffer an unspeakably cruel death). Then there are ghost lines, so named because they've drifted away from the vessels that set them; the lines float like tangled spirit worlds that swallow marine life that tries to pass through them.

Alex Cornelissen is the director of Sea Shepherd Galápagos and Sea Shepherd Europe. A quick-witted Dutchman, Cornelissen is no less defiant than his shipmate Watson. His monkish hair, quiet voice, and featherweight frame provide the perfect foil for the usual pirate image. Like Watson, he has captained ships to the ends of the Earth—from Canada to South Africa to Antarctica. Cornelissen and Watson's then-wife, Allison, once infiltrated an inlet in Taiji, Japan, called the Killing Cove, where 2,300 dolphins and small whales are slaughtered every year. A photograph of the inlet shows fishermen in a wooden boat afloat on a sea of blood. The two eco-pirates sneaked in, severed the nets, and freed more than a dozen dolphins before they were assaulted by fishermen and arrested. Author Peter Heller writes in *The Whale Warriors* that the Japanese government allows fishermen a quota of more than 20,000 dolphins and other cetaceans every year. The meat is sold to school lunch programs and grocery stores, even though it contains alarming levels of mercury.

Cornelissen became head of Sea Shepherd Galápagos in 2007 when the director, Sean O'Hearn, was deported from Ecuador under false pretenses, not long after he helped the Environmental Police legally bust a major shark fin mafia based on the Ecuadorian mainland. Sea Shepherd already had a working relationship with the park. Back in 2000 the organization signed a five-year contract with the park to provide a new patrol vessel, the *Sirenian*, which was greatly needed in the barely patrolled waters. In the next few years the *Sirenian* chased and caught hundreds of poachers, and confiscated their illegal catches and their vessels. The ship, formerly named the *Edward Abbey*, after the famous desert activist and author, infused the park administrators with new energy and verve. When the contract ended Sea Shepherd donated the vessel to the park to continue patrolling the archipelago. After five years on the high seas, the ship was badly in need of repairs. In 2006 the World Wildlife Fund entered into a partnership with Sea Shepherd and donated $400,000 to overhaul and refit the former Coast Guard cutter. It was then renamed the *Yoshka* by request of the donor.

One of my goals while I am living in the Galápagos is to volunteer on the *Yoshka* for a couple of weeks, even if it means being a galley wench, slicing and dicing and baking bread. But Cornelissen tells me this is unlikely. Despite its recent overhaul, the ship is still anchored in the bay. He's seen it go out only two or three times because the park doesn't have enough crew to run it. That's one reason, he confesses, that even he hasn't been out on the boat. It's a letdown. Like the Sea Shepherd volunteers, I want to sail on the front line of pure adrenaline, to feel the drama of running down poachers, to understand directly why this ocean is in danger.

I settle instead for talking with Cornelissen in Sea Shepherd's office, where a huge pirate flag hangs on the wall behind his desk. Another wall displays hooks confiscated from illegal longlines—a stark reminder of Sea Shepherd's mission. A longline can range from 1 to 62 miles in length and is attached to a plastic or Styrofoam buoy. Every 100 feet a secondary line extends about 16 feet deeper. The lines are hooked and baited with fish, sea lion meat, or chopped-up dolphin and set adrift for up to 24 hours to attract sharks. Worldwide, more than 8,000 tons of shark fins are processed every year. Fins equal about 4 percent of a shark's body weight. Do the math and it means that about 200,000 pounds of shark are dumped into the ocean every year. Shark fin soup, a delicacy in Asia, can cost between $50 and $400 for a *single bowl.*

The shark is an apex predator at the top of the food chain, a key species in maintaining a diverse and healthy marine ecosystem. Sharks keep fish populations stable; they also scavenge diseased and dying species. Removing them from the oceans can throw an entire ecosystem into chaos. Sharks, you could say, are more fragile than they are ferocious. They reproduce slowly, reaching sexual maturity at about 15 years and giving birth every year to a single pup. Yet every year humans kill more than a hundred million sharks. The International Union for Conservation of Nature (IUCN) now lists 18 species of shark as endangered. In contrast, marine biologists are quick to note that more people are killed by lightning strikes on golf courses each year than by shark attacks.

Cornelissen and I settle into a couple of overstuffed chairs to talk. A cool breeze blows in from the bay, where pelicans perch on the *Yoshka* looking for fish. What follows next is a Q & A. Call it oral history.

Q: Was there anything in your young life that inspired you to be so passionate about nature?
A: No. I grew up in the doom, no-future period—1980s, new wave, and punk rock. *There's no future. We're all going to die because of the bomb.* I used to live next to the NATO headquarters in Europe, in the city next to the town where I was born, so I grew up with the thought that if the bomb ever strikes I'll be the first to go.

Obviously that's changed now, but no, I was never really involved in any environmental activism until I met with Sea Shepherd. I was always interested in [the environment], but I never really found anything that I thought was worth supporting. I've always said: *I don't want to give to an organization because I don't know what's going to happen to that money.* Then, when I heard Paul Watson give a speech in Amsterdam in 2001, I was just blown away. The ship left with a friend of mine onboard and made a documentary for Dutch television. When my friend came back I decided to pack my stuff and go, so I sold my apartment, quit my job, and took a sabbatical. After I came back that year, I just couldn't live at home anymore so I [left again] and have been doing this ever since.

Q: What does being a captain involve?
A: As a captain, it involves me ... things like the seal campaign. We were just up in northeastern Canada documenting the Canadian seal hunt. Our goal is to shut down the Canadian seal hunt, obviously. We've been trying to do that since the 1970s. This year was very successful because we [drew] international attention to the slaughter. The Canadian government completely overreacted, which obviously gave us a lot more publicity. We got rammed by the Canadian Coast Guard twice. When we went to the French islands of St.-Pierre and Miquelon we got attacked by a mob of angry fishermen with axes who cut our lines and set us adrift, almost hitting the rocks. We could barely escape. If we had started our main engines like ten seconds later, we would have rammed the rocks and created an oil spill there, obviously not our fault. During the campaign we were eventually boarded by the RCMP [Royal Canadian Mounted Police], a tactical unit with machine guns and shotguns. They attacked our vessel with full squad gear. We were held at gunpoint. The entire crew was arrested, then later released. Only the first officer and I were charged with violation of the Seal Protection Act that basically states that you're not allowed to come

within half a nautical mile of someone killing a seal. We were arrested and almost killed for taking pictures of someone skinning a seal alive. I still don't get it. *We were actually arrested at gunpoint for taking pictures of someone skinning a seal.*

Q: Is it hard not to get emotionally involved?
A: You can but not while you're [engaged in an action]. You can have nightmares later. I can tell you I do. But while you're standing on the ice with your feet in the blood of a seal, and there are seal carcasses all around you and people are smoking cigarettes and making masturbation motions while they have their dicks out of their pants then, yeah. These people really have no reason to be there. I'm talking purely about the sailors now. A lot of them actually say that they go out to the ice to kill seals because it's a good reason to get away from the wife and kids and be out with their friends, drink some beers, and kill something. It's a blood sport. There's no other description for it.

Q: Have you ever feared for your life?
A: Not while in Sea Shepherd, no.

Q: Not even in Canada recently?
A: No, not at all. It's just a strange experience when your bridge wind doors get kicked in by 12 RCMP cops with machine guns and shotguns and tell you to lie down on the floor before they shoot. Yeah, that's a strange experience. We sort of expected it to happen. So before they came we raised our arms and faced the door and said, like, *Okay, here we go.* Then they kicked the door in. You know what to expect and you just have to play the game.

Q: What is Sea Shepherd's main objective in the Galápagos?
A: We want to stop poaching in Galápagos. We want to make sure that

the Galápagos has not been raped and plundered like the rest of the world because this is one of the few pristine areas in the world and obviously we want to keep it that way. So we do whatever we can. I think of cooperation with the Galápagos National Park, cooperation with the Environmental Police, and cooperation with nongovernmental organizations (NGOs) to make this a better place.

Q: You go out with the blessing of the Galápagos National Park?
A: We haven't actually been on patrol with them because our vessel, the one that we donated to the park, the *Yoshka,* formerly known as the *Sirenian*, hasn't been running for two years because it needed repairs. It's anchored right now. I've only seen it go out two or three times since it's been back in operation because the park only has a crew for one and a half boats. But they've got three boats so they've got a logistics problem. They have too many boats and not enough people.

Q: You said Sea Shepherd donated this boat to the park?
A: Yes. We bought it in 1990 I believe. It's a former U.S. Coast Guard cutter; we gave it to the park in 2000 on a five-year contract/lease, and after those five years we donated the vessel to the park. It's now registered in Ecuador, previously registered under Sea Shepherd's name. Now it's a park boat but we still have the right to participate in patrols.

Q: Have you gone out with the Environmental Police?
A: I myself have not gone on patrol with the Environmental Police but we are in the process. We actually just bought six sniff dogs, which are going to be used to find marine products like sea cucumbers, shark fins, sea lion penises, seahorses. You'd be surprised. We bought some dogs in Colombia and they've got to be trained. They're like drug dogs only these are specialized in [sniffing out] marine products. They're Labradors. They're really cute. The problem is that the dogs have got

to be trained, and you're dealing with the incredible slowness of the Ecuadorian system.

Q: Each dog has a handler?
A: Yeah, a handler who has to like dogs and they have to be able to train the dogs because these dogs need constant training. It's not like you train them and then they go out and find sea cucumbers. You have to constantly train them with like scents, … so it's a constant process. If you don't do it properly these dogs are going to be useless in a few months. We're starting a Sniff Dog Unit for the Environmental Police. They've never had that. They've had two sniff dogs before, which are here in Santa Cruz. Those are really, really good dogs. They came from the States. I believe they spent $25,000 per dog. They were already trained when they came here. According to the director of Wild Aid in the Galápagos, Godfrey Merlen, the dogs apparently were so good that a bounty of $50,000 was placed on each canine's head.

In February 2009 Sea Shepherd's new sniffer dogs had been trained and were ready to work.

Q: How effective has Sea Shepherd's longline campaign been in the Galápagos?
A: We would like to do more against long-lining. We certainly have the knowledge. Every single time we've been here with the *Farley Mowat*— and I've come back here five times since 2002—we've spent about a month in the Galápagos. We're not allowed to go on patrol [alone] because we don't have any patrol rights, but we do it anyway. Every single time we've done so we've caught poachers.

Q: What have you found?
A: In 2002 we caught a long-liner inside the park, freed a sea turtle, released several sharks, and pulled in about 10 to 15 nautical miles

of longline. Then we came back in 2004. We caught a Costa Rican long-liner, an Ecuadorian long-liner, and an Ecuadorian tuna boat. The Ecuadorian tuna boat eventually just left. They were about our size and they didn't really want to wait around for the park authorities to show up. The two long-liners eventually got taken in—one by the *Sirenian* and one by the *Guadalupe River,* one of the park's two other patrol boats—so it was a very successful campaign for the park and for us. At that point we were really in close cooperation with the Galápagos National Park and they were more than happy to use our assistance.

Q: Did things change in 2005 when the new park director, Raquel Molina, came onboard?

A: No, she wasn't the problem. It was other people in the park service. It's very political. People here consider this their turf, and any foreign organization interfering with their business is not welcome, even though we certainly have the expertise and knowledge to stop poaching in the park. I know exactly where the poachers are. If you give me a ship right now I'll go out and tomorrow morning I'll catch a longline, I can guarantee you. Why? Because I've done it every single time I've been here. The last time we went to Isabela we had some mechanical difficulties and ended up in an area where poaching was going on. Early in the morning we found several speedboats deploying longlines. So we stopped and started pulling in the longlines. We sent one of our speedboats after them—took pictures, took video, recorded everything—and then later handed the evidence over to the park. Their reaction was, *Well, you shouldn't have taken the longline in. You should have called us*—instead of saying thank-you. I got a little pissed off about that.

Q: What are the worst things you've seen as a result of longlines?

A: Probably the worst I've seen in the Galápagos was a six-foot manta ray with a longline hook caught in its back. He had a piece of flesh about eight to ten inches just ripped off his back and he was severely injured. We took the hook out but I don't know if he actually made it. This makes you very sad because that line is in the water to catch tuna or shark and the by-catch is really bad. In 2002 we found a sea turtle. When turtles see longline buoys they think it's another turtle so they start humping the buoys and then they get their flippers entangled in the line and they drown. Fortunately, we were able to save that one, but on any given day there are a number of turtles that die because of these longline buoys. You also see by-catch in the amount of seabirds getting killed, like albatross. Worldwide, longlines are the biggest problem of seabird mortality, and now pretty much all the species of albatross worldwide are endangered because of long-lining.

Q: For such a huge marine reserve, it seems the issue of patrolling and management is impossible.
A: It is difficult but you've got to realize they're not fishing or poaching in the entire Galápagos Marine Reserve. There are a certain number of hot spots where we know that poaching takes place on any given day, and that's where the focus should be.

For instance, to the southwest of Isabela, there is an area where on any given day hundreds of miles of longlines have been put out. There should be a permanent patrol there. Then you've got the area basically around Wolf and Darwin. We know there is poaching going on there. Fortunately, the *Tiburón Martillo** is now anchored here in the harbor. That's a barge they're going to tow up to Darwin and Wolf Islands then

---

*In October 2008 the park finally added the Tiburón Martillo to its fleet to serve as a permanent base off Wolf Island, to monitor poaching in this shark mecca, and to conduct scientific research. The base station was made possible by donations from Wild Aid, World Wildlife Fund, Conservation International, Ecoventura, and Sea Shepherd. Peter Witmer, owner of the Aggressor I & II scuba-diving fleet in the Galápagos, donated the Tiburón Martillo to the park. It took eight years for this to become a reality.

permanently anchor to the sea bank. Put a couple Zodiacs next to it and it will be a permanent floating park station. That's really good but they're been working on that since 2000.

Q: Why so long?

A: There have been problems with permits, and the [Ecuadorian] Navy is giving us a hard time, and that's one of the problems we're facing. The Navy is really not doing what they should be doing. For every tourist that comes into Galápagos, $10 of the $100 park entrance fee goes to the Navy. Why? So the Navy can patrol the national park. I haven't seen any Navy boats around here lately—or at any time. The only time I've ever seen the Navy is when Sea Shepherd's boat was at anchor here and the Navy was next to us watching us to make sure we didn't go out patrolling. That's what the Navy does. Now, the Navy should be patrolling [the marine reserve]. If they would keep their patrol boats around the archipelago and keep patrolling the area, there would be a far lower number of poachers.

Q: Why isn't there more monitoring in the marine reserve?

A: Lack of funds, you could say, but I don't think that's the case. I think they're not because—and this is unfortunately the case—there is a lot of corruption within the Navy. A lot of people within the Navy only have a two-year contract, and in those two years they want to cash in. I'm talking about the officers. I'm talking about the middle staff. They accept bribes. We know for a fact—but we don't have evidence to prove it—that sometimes the Navy even uses Navy boats to transport poached products to the mainland. I think everybody in Galápagos knows it—it's the word on the street, the buzz on the street. Every NGO knows it, but nobody talks about it.

Q: What kinds of products are they smuggling?

A: Shark fins, mostly, sea cucumbers, and sometimes when the park catches a poacher the Navy will come in and release that poacher again. At the moment, there is a fishing boat near Isabela that's been arrested by the park and the Environmental Police. The Navy was trying to get them released because clearly someone within the Navy had accepted some money. Fortunately, both the police and the park were present and a lot of people know about it, so they cannot let them go at this point because then it would clearly be a matter of corruption.

Q: What would it take to successfully protect the marine reserve?
A: You need a fleet of 20 ships to patrol the area constantly, and you need the Navy to cooperate as well, because it is a big area. If you do that for a number of years, then eventually the poachers are going to disappear because they'll realize there is no money to be gained. If they're caught every single time they come in here what's the point? It's going to cost them more. If it's costing them more money than they're making, they'll go somewhere else.

Q: Are Darwin and Wolf Islands prime targets because of the number of sharks?
A: Yeah, enormous amounts of sharks. Not only that, it's a very remote area of the park, and there is not that much control going on up there, so people have got free play. The only boats that come there are tour boats, dive operators. What you see in Darwin and Wolf is they're mostly Costa Rican long-liners. What you're seeing southwest of Isabela is mostly Taiwanese long-liners. What you're seeing northeast of San Cristóbal is mostly Ecuadorians. Those are with the bigger Taiwanese long-liners. What they're doing is paying either local or Ecuadorian long-liners to go within the park, catch the tuna or the shark, then bring it outside the park boundaries. Then they off-load and pay them.

Q: They're using middlemen?

A: Yeah. They're staying outside the limits [of the marine reserve], and you can only catch them when they're off-loading the fish, which hardly ever happens.

Q: Sea Shepherd launched an investigation into the indictment of Villamil's mayor, Pablo Gordillo, for cutting down protected mangroves. What's Sea Shepherd's experience with him?

A: He's a piece of work, the mayor. I think he's probably the biggest menace to the Galápagos. What he's been doing—and a lot of what I'm telling you is definitely confirmed. He has cut the mangroves. We have that on video. He's been prosecuted for that. He has bribed the judge to get away with it. We're still investigating that matter. That [case is] not over.

Q: Who videotaped him?

A: It was Henry Segovia.

Q: They have a long-term feud, don't they?

A: Yes. Henry had a bar on the beach for which he has been given permits every year. Immediately after he videotaped the cutting of the mangrove his permit was denied, and since then the mayor has done everything to make his life impossible. ... [The mayor] is basically doing whatever he wants. He is really abusing his power.

Q: The mayor took me to the site and told me the mangroves he cut were just branches and dead wood.

A: It was not dead wood; there was nothing dying there. He cut the mangroves. There is no question about it. We have video evidence to prove it. But somehow judges got bought and nothing happened, and that's exactly the problem. This country is corrupt. People put out enough money and nothing happens. If you have enough money you can get away with anything.

Q: What was Sea Shepherd's role in fighting the company Planktos in 2007, when it planned to dump iron dust into the ocean just off the Galápagos?

A: We left Melbourne, Australia, and I was captain on that trip, with a stop in Galápagos and then through the Panama Canal up to Iceland to interfere with Icelandic whaling. While we were in Galápagos, we found out about the Planktos plan to come to Galápagos to dump iron in the ocean.* It was quack science. There is clearly no scientific background for what they're doing. What Planktos was doing was cashing in on a new trend. While we were sitting here [in the Galápagos] we found out about it, and we also heard that Iceland was reconsidering its whaling that year so we didn't really know what to do. We had two possibilities. We could either proceed on to Iceland, do a campaign [against something] that might not happen, or stay here while we had the perfect opportunity and the perfect location to stop Planktos. Eventually, Paul [Watson] and I and some other people had a discussion, and we decided to stay here in Galápagos just to wait for Planktos. Our presence here deterred Planktos from coming. They said they had problems loading their equipment onboard in Florida, but we know that they were scared by the announcement that we would physically stop them—and we would have stopped them physically because there is absolutely no need for this sort of experimenting in Galápagos.

Q: The Environmental Protection Agency in the United States was against their operations and told them the iron dust couldn't be transported on a U.S. flag vessel.

A: The fact that the U.S. government was grilling them for their scam research pretty much gave us the right to interfere with them, and that's

---

*The company claimed that dumping iron dust into the ocean would cause phytoplankton to flourish and absorb harmful greenhouse gases, such as carbon dioxide.

why we posted it on our website.* That's why we publicly announced that "we will physically stop you if you come here" to the Galápagos.

Q: I read that Planktos went out of business and the CEO resigned.
A: I think that Sea Shepherd had a very important role in that; we put them out of business. They turned up in Bermuda, which was where the *Farley Mowat* was docked, and we immediately staged a protest. Our crew was there. We put it on our website, and so they left Bermuda and ended up in the Azores, but that was the third time they had been stopped by our mere presence. Eventually that led to a collapse of their stock and to the resignation of their CEO and basically the end of the company.

Q: So it was Sea Shepherd that stopped Planktos in every case?
A: Not intentionally, but we just happened to be in the right spot at the right time. What makes me say that is that Planktos themselves actually said that due to interference of NGOs, Planktos was put out of business. They actually gave us the credit themselves and, of course, we'd be more than happy to take that credit. Yeah, you could say that we put Planktos out of business and I'm very proud of that.

Q: Tell me about the flags you've painted on the *Farley Mowat* from boats that were rammed and sunk.
A: The first one was the *Sierra,* the one that was rammed in the Lisbon Harbor by Paul [Watson] back in 1979. The *Sierra* was a pilot whaler

---

*At a symposium held at Woods Hole Oceanographic Institution in Massachusetts that same year, the world's most prominent marine biologists and climatologists challenged Planktos's claims and offered research that proved iron seeding is scientifically unsound. The scientists proved that too much phytoplankton sucks up oxygen and suffocates marine life. Planktos had received funding in part from "carbon offset" sales, and planned to launch its project about 350 miles from the nutrient-rich waters of the Galápagos.

responsible for the killing of—I believe it was 30,000 whales. Paul's mission was to shut down the *Sierra,* and he chased her all over the Atlantic and finally found her in the harbor of Lisbon, Portugal. Then he rammed it amid ships, turned around, and rammed it again. He was charged by the port authorities in Lisbon for poor seamanship and for poor handling of a vessel, and he turned around and said, *It wasn't poor handling. I intended to hit her and I hit her exactly where I wanted, so why don't you show me the charge?* But there were no charges, of course. It was a pirate whaling corporation. The ship was run by a Norwegian crew owned by a Japanese company, so all the whale meat was eventually transported back to Japan. Then there are two [flags for] pirate whalers, which were somehow owned by a Spanish company. Then there are two Icelandic whaling ships back in 1986 when Sea Shepherd shut down Icelandic whaling by sinking half their fleet and destroying the whale meat processing plant in Reykjavik. There are flags for three Norwegian vessels, one Japanese drift-netter, and two Taiwanese drift-netters.

Q: Sea Shepherd sank half of the whaling fleet in Iceland?
A: Have you heard of Rod Coronado? He was 18 at the time. Rod Coronado and another guy went to Iceland and they sank two ships. They were going to sink a third one, but there was someone sleeping onboard and obviously we do not want anybody to be injured, so they left that ship alone. They had four whaling ships back then. Two of them were sunk in the harbor. Then [the activists] moved on to a whale meat processing factory and pretty much trashed the place—destroyed it completely. Our main goal is to get the insurance fees raised on whaling vessels. They actually have to pay for war insurance.

Q: Why do whaling ships have to buy war insurance?
A: Their insurance fees are the same as countries that are at war. ... They're so high that a lot of companies can't afford to pay their insurance

fees anymore. So they're either uninsured, which means that if you attacked them and sank them they'd pay such enormous amounts of money that they'd be going out of business. That's what happened in Iceland. The insurance fees got raised so much they decided to stop whaling.

Q: So Sea Shepherd shut down Icelandic whaling?
A: Yes, in 1986. After the *Sierra* got rammed in Lisbon Harbor they completely patched her up. The same day they finished repairing her, two Sea Shepherd—you could say—commandos went onboard and sank her again, and that was it. This is what happens when a ship is rammed: They ram the power blocks. Those are the blocks that pull in the nets. They ram that part of the ship so they're no longer capable of putting out or pulling in the nets. The ones that are sunk are all sunk in port. We don't sink ships at sea. People say, *Are you going to sink some ships again?* When we went to Antarctica, *Are you going to sink a Japanese ship?* "No, of course not. We don't want to jeopardize people's lives," and especially in an environment like that. It's way too dangerous. But we don't shy away from sinking them at dock. As far as we're concerned they're illegal vessels.*

Q: You're willing to put your life on the line to protect natural resources?
A: Absolutely. What better way to risk your life than that? I mean, is it better than risking your life for a bunch of oil, like they're doing in Iraq

---

In 1986 the International Whaling Commission (IWC) banned commercial whaling, but Japan kills about a thousand whales a year under a scientific whaling program that Japanese authorities claim provides important data on whale distributions, populations, and feeding habits in the iceberg seas near Antarctica. Sea Shepherd and other groups—animal rights groups—condemn this as a ploy to continue commercial whaling. But the IWC lacks the ability to enforce this ban. Enter Sea Shepherd, whose mission is to enforce these international conservation laws on the high seas.

right now? Those people are putting their lives on the line to protect oil and money. Here we're protecting the marine environment, and I think that's a noble cause. Of course I'm willing to risk my life for that.

Q: Has anyone with Sea Shepherd ever died?
A: Nobody has ever been injured—not on our side, not on the opponent side, except when you want to think about a broken finger or stuff like—that but no serious injuries. No injuries in the line of duty—none—and we're proud of that. Thirty-one years without injuries.

Q: How do you feel about sportfishing?
A: I hate it, absolutely hate it. I used to go sportfishing when I was kid and my dad used to take me to a pond. You throw your line, you catch the fish because, *Ahh, fish don't feel the pain,* but they do. You're hooking a fish by its mouth. You're yanking it out of the water. He's suffocating and then people have the audacity to say, *Oh, fish don't feel pain.* Of course they feel pain. It's a horrible way of fishing, and what you're seeing with the sportfishing that takes place at sea, for instance, they're catching marlin and tuna and shark and then they're doing catch and release, which is absolutely ridiculous. For instance, if you pull a marlin in he'll fight so badly to get away that the fish will burn up all its reserves. So when that fish is released it drowns because it doesn't have the strength to stay alive anymore. It's given up basically. The moment you bring it on board, that fish has given up its life.

Q: But there's a lot of pressure to legalize sportfishing in the Galápagos.
A: That should never happen. If that happens you can just wait for the destruction of the Galápagos. What you're seeing is that a lot of the fishermen are taking tourists out on their boats, and they allow them to go sportfishing with them. I think it's ridiculous. It's like a subsidized hunting trip—like rhino hunting in Africa. If you have enough money,

you can do that. But why would people go out on the boat to catch these beautiful fish? Why? To make themselves feel big. *I was able to pull this beautiful fish out of the water and in doing so I killed it.* *"Good job."* There used to be boat called the *Galápagos Shark*. They actually did sportfishing trips, and we followed them in 2004 with one of our Zodiacs. They saw us following them, turned around, and started chasing us, so we went away. Then we started following them again. Eventually they called in another boat and tried to cut us off, and we had a cat-and-mouse game with the two sportfishing boats. Later on it turned out that they'd filed a complaint with the mayor and the Navy about our interference. They didn't want anybody watching what they were doing because they were clearly fishing illegally.*

Q: If sportfishing were legalized here, would Sea Shepherd take action?
A: If it's legal there is nothing we can do about it. That's the problem. Sea Shepherd operates within the boundaries of the law, and even though we really seek out the edge of the law, we will never break it. We have to comply with regulations. We're here in Ecuador as a registered Ecuadorian nonprofit [group], but we have to follow the Ecuadorian law. We can't just say, *Oh yeah, we don't agree with that, therefore we're*

The Navy never prosecuted the boat owners or anyone else for illegal activities in the marine reserve. In fact, the mayor of Puerto Baquerizo Moreno (the provincial capital of the Galápagos on San Cristóbal Island), and the mayor of Puerto Ayora on Santa Cruz Island have held sportfishing tournaments in the marine reserve over the last few years. In June 2006 the Tenth Annual Father & Son Billfish Classic Tournament was advertised on the Internet to promote sportfishing in the Galápagos. A Galápagos National Park ranger who flew over a sportfishing boat called the Albermar to investigate was sued by a group on board. The suit alleged that the pilot had endangered the lives of those on board by flying too low over the boat. The pilot replied, "I know the rules and never would risk my life nor others." His report stated that the boat was clearly sportfishing illegally within the marine reserve. Park Director Raquel Molina stated, "The group, which included a naval officer, were sportfishing and threatened to take away the park pilot's license and kill the warden in charge of the flight."

going to break the law. No, we don't do that. We might not agree with it but unfortunately if that's the case, if that's the new law, then, yeah, we're going to have to stick to it. But if there is any way that we can help politically to stop such a law from happening of course we'll do whatever we can.

Q: What was Sea Shepherd's working relationship with Galápagos National Park Director Raquel Molina?
A: I think Raquel was probably one of the better park directors that Galápagos has ever known.* I can guarantee you that she was not corrupt in any way and she had the best intentions for the marine reserve. I think her problem was that she didn't play the political game, which I think is a good quality but unfortunately in Galápagos *that's* eventually going to work against you. We were quite happy with her. We had very good cooperation with Raquel.

Q: Do you think she got fired because she refused to give the tour ship *Pinta* a permit?
A: Yes, I think that's the case. The [environment] minister was saying that Raquel was fired because she was insubordinate. But at the same time the minister allowed for that permit to be handed out, and Raquel turned around and stopped it, and she had all the right to do so. She was the park director and the minister didn't like that and fired her.**

Raquel Molina was the first ever female park director and the 13th within only a few years.

**Ecuador's leading newspapers reported that Molina was fired for not allowing Roque Sevilla, one of the most powerful tour operators in the Galápagos and executive president of Metropolitan Touring, more than one permit to run a single ship, the Pinta. I tried to reach Molina but she would not respond. Metropolitan's director of operations in Puerto Ayora, David Balfour, told me that according to Metropolitan's lawyers, the company's actions were legal. The Pinta began operating right after the park director was fired.*

Q: What happened to Sean O'Hearn? Were you here when he was Sea Shepherd's first director here?

A: Yes. I've known Sean since I started working for Sea Shepherd. Sean actually started working with Sea Shepherd in 2000 before I came along. He was one of the first people here in the Galápagos who was catching poachers along with the park. [Once] when the park was seizing the poaching boat, *Maria Canella II,** the Navy guy on board got a call from a high-ranking official and was ordered to abandon the ship immediately. The owner of the fishing vessel had some ties, so that fishing boat got released. Sean was actually aboard that ship doing the inspection when they got the call: *Leave everything the way it is and abandon the ship. The people are to be released.* That thing got blown up in the media.

Q: What was confiscated in the sting operation in 2007?

A: Sean went to the mainland and that's when he confiscated 90,000 sea cucumbers and 20,000 shark fins. He didn't personally confiscate them; the Environmental Police did, based on the information Sean gave them.

Q: Did that land him in trouble again?

A: No, that didn't land him in trouble. While that happened [in July 2007] we were here in the park patrolling and we found a longline that was about 30 miles long. While we were pulling in the longline Sean was making the bust on the mainland. We were all over the place—and

---

* *The* María Canella II *was seized by the park patrol boat* Sirenian *for illegal longline fishing near Wolf Island. The ship's hold contained numerous dead sharks and 1,047 shark fins. In May 2002, at four o'clock in the morning, the boat escaped and fled to Costa Rica, according to the port authority captain in Puerto Ayora. Sean O'Hearn was deported but returned to the Galápagos a few years later and became the director of Sea Shepherd Galápagos.*

in the news—and obviously our actions started to draw some interest from the people on the other side of the game, the mafia.

Q: What do you mean by the mafia?
A: Fishing mafia and fishing fleets, the persons normally involved in fishing illegally. They're very powerful because it's a large group in a country like Ecuador. There are a lot of fishermen and they have families, and these people are all voting. [Voting is mandatory in Ecuador.] By getting the fisheries behind you you've got a lot of votes. So obviously presidents are trying to keep these people happy because that would perpetuate their reign in government. Then Sean was involved in yet another sting—probably the biggest operation ever on the South American continent—which involved shark fins. We don't know exactly how many shark fins were confiscated but they went into a house in Manta, and in that house there was an estimate of between three and five tons of shark fins and dried shark fins. These shark fins were confiscated by the police, but while they were doing the operation again a call came from some high-ranking official and everything was returned to the fishermen.

Q: They returned tons of poached fins to the fishermen?
A: Yeah. The president said, *Return everything to the poor fishermen.*

Q: Isn't that illegal?
A: You would imagine so, wouldn't you? There was definitely someone on the take. As for shark fin mafia, it's a worldwide problem. There is a ton of money to be made in this industry. You've got prostitution, you've got drugs, and you've got shark fins, and the shark fin business is rising and there's a lot of money there. I heard yesterday that 20 percent of the policemen here in Galápagos are actually here because they can no longer be on the mainland because there's a price on their

heads—because there are contracts out on them. So they are sent here as a security measure for them personally. There is a lot of organized crime on the mainland. People rob banks all the time. People rob grocery stores, so there are people with shotguns and machine guns and pistols walking around all over the place. As a policeman you would definitely be a potential target, and apparently 20 percent of all the policemen here in Galápagos are—or have been—potential targets. What I heard later is that one of the president's advisers called the president and said that some gringo was kicking in the doors of poor fishermen and stealing their fish. [Sean] doesn't have the authorization to begin with. He was there as an observer. He was an informer. He was channeling and informing ... to the police, and with that information the police made the arrest. But it's not like Sean was kicking the doors in. He was merely there to make sure that everything that was confiscated was burned.*

Q: He was denied habeas corpus to stop the deportation?
A: Yeah. That was denied, which is not legal. Then Sean's wife came on the newscast, saying that Sean has an Ecuadorian residency. His wife is Galapagueña and he has a two-year-old daughter who's Galapagueña. So the president was deporting someone who had the right to be here. Her presence in the media created a lot of attention and people were outraged about it. Sean was actually on the way to the airport when the order came to release him. At that point the situation for Sean was very dangerous. Some people in the shark fin mafia where the big [bust] took place in Manta probably put a price on Sean's head. His face was in all the newspapers, all the TV shows, so it's pretty easy to pinpoint a person, and unfortunately that was Sean. At that point he decided it was no longer safe to be in Ecuador, so he fled to the United States. [He

---

*Photographs show huge piles of fins being incinerated as O'Hearn and the Environmental Police look on.

came back] for four or five days and found out that people were still looking for him so he decided to leave Ecuador permanently.

Q: So you took his place?
A: Yes. I was just here in the right spot at the right time.

Q: Why do the fishermen here have so much political clout?
A: The fishing industry, not only here but worldwide, is an industry that for some reason, no government wants to mess with.* They're always trying to please the fishermen.

Q: What's your personal affinity for sharks?
A: I love sharks. I've always loved the oceans but specifically sharks. I think sharks are possibly the coolest animals on the planet. I mean just the fact that they haven't changed for tens of millions of years—and why?—because they're perfect. They're the perfect animal in the ocean. They've been on top of the food chain forever, and you have to respect such an animal that's so close to perfection. That's why I love sharks so much.

Q: When was the first time you saw a shark up close?
A: Strangely enough, the first time I saw a shark was after I joined Sea Shepherd. It must have been in the water while I was diving. I think the first time was near Tonga in the Pacific—it's just south of Fiji, south of French Polynesia—and I saw it from a distance so I didn't really get too close. But then I saw sharks near Montello, but the biggest concentration of sharks I was fortunate enough to see was here near Darwin where we were diving. I was seeing an enormous number

---

*Sea Shepherd says the Ecuadorian government has a severe case of homopechephobia—a political fear of fishermen.

of whale sharks. I swam in schools of a hundred-plus hammerheads. But the ones that really impressed me the most were the Galápagos sharks, because you don't mess with those. Those are like the little brother of gray-whites. They're closely related to gray-whites, and they're very beefy, big, mean sharks. They look mean but they're just fantastic.

Q: Have you ever seen a shark with a longline hook in its mouth?
A: I saw a hammerhead shark with a hook in its mouth while I was diving off Darwin, and [at the time] there was a fishing boat right on top of Darwin Arch, which is illegal.

Q: Do sharks survive if they're hooked?
A: Yeah. As long as the line snaps they can totally survive. Eventually it will come out but if you're in the water and a shark has been freshly caught on that line and released he gets very aggressive. They get very agitated. While we were diving we saw the hammerheads, fins down, and cruising at high speed.

Q: If a shark bleeds does that attract other sharks?
A: No, they don't really bleed that much. They just wear out. Sharks are really tough, so [getting hooked] really doesn't harm them too much. We've actually had sharks, like three- to four-foot sharks, that we pull on board the *Farley Mowat*. We hoist him out of the water, put him on the deck, clip the hook out with bolt cutters, then pierce the hook out of the shark, and chuck him back into the water and, bang, off he goes. They're very, very strong. Sharks, even when [poachers] cut off their fins, [struggle] in the water while they're still alive; that's how strong they are. That's another reason why I'm so full of respect for sharks.

Q: Do you have special training for passive resistance?
A: No. With Sea Shepherd, we instruct the new crew members about

what's going to happen, what can happen, what you shouldn't do, and what you absolutely have to do. So before we got boarded [in Canada] we had a meeting and said, *If we do get boarded, whatever you do don't resist. Don't pull out a knife or anything like that.* Because they will shoot you, and the last thing we want is someone getting hurt over this. There are enough animals being killed, and the last thing we want is crew members getting killed.

Q: When was the first time you went on the *Farley Mowat*?
A: First time was on July 11, 2002—actually just six years ago here in Galápagos. I was chief cook. I started with vegetarian and vegan and meat because we had all three aboard, and then we slowly changed to vegetarian and eventually became vegan. All Sea Shepherd vessels are now vegan.

Q: I saw empty rum bottles in the recycling bins aboard the *Farley Mowat*. I take it it's okay to drink on board?
A: Of course. We're pirates. We do have obviously some alcohol on board. You can't expect people to be completely devoid of any alcoholic beverage. But while we're on campaign I personally always want the ship to be dry, and I don't want to be at sea with a drunkard. I don't mind people after a day's work having a drink. It's really no problem. We're not the Marines. We're a volunteer organization and you have to treat people like that. But, of course, intoxicated behavior will never occur because people are there for a reason, not to have a party. If you want to party, go join Greenpeace. But I was only a cook for about six months. So after the Antarctic campaign, which ended in January/February, I joined the ship on the bridge and was second mate. Then we did about one year in Seattle dry dock and some repairs. When we left Seattle I was first mate and I was first mate for about two and a half years until I became captain.

Q: When were you the captain of the *Farley Mowat?*

A: I was the captain of the *Farley Mowat* when we left South Africa [in 2006]. That was my first trip. We were under detention in South Africa. We were not allowed to leave because they were asking us for commercial payments, which we didn't have. So my first action as a captain was to leave Cape Town harbor without permission. We did that in the middle of the night—at three o'clock in the morning—and they didn't find out we were gone until eight o'clock in the morning, and that's when they sent the [South African] Navy after us. But of course by then we were well outside the 12-mile limit.

Q: You're pretty fearless.

A: It's part of the job. You were asking me if I was ever scared. That was probably one time that I was scared because if they'd caught us I would have been in serious trouble. I don't want to go to jail in South Africa. We were very worried about it. We actually were thinking about taking the ship out with just the chief engineer and myself and then having the rest of the crew come out on a sailing boat and pick them up at sea, just in case we did get caught—then only he and I would go to jail. But eventually we decided just to make a run for it and it paid off well. Then we went from there to Fremantle-Perth, Australia, and then I commanded the ship from Perth to Melbourne. [After that] I flew to the Caribbean to look out for a new ship; that ship didn't turn out to be very good, so I flew on to Scotland where I found the *Robert Hunter*; got her ready for departure and then I was the first captain of the *Robert Hunter*. Then took her from Scotland to Punta Arenas, Chile, and then through the Strait of Magellan to Antarctica where we did the campaign. ... [T]hen Paul [Watson] and I swapped ships about a week before we came back to port. Then I did a trip from Melbourne to Bermuda, stopping here in Galápagos, and then the last campaign was the seal campaign in Canada. Those are the campaigns I did as captain, and they will also be the last of the *Farley Mowat*. *Farley Mowat* is under detention at the moment in Canada. The Canadian government has her dockside with, I think, 24-hour security guards. They're watching the ship. The ship is slowly deteriorating. But what they don't know is that we, quite frankly, don't want the ship anymore; the ship has been retired. We were like five days away from retiring the vessel and everybody [in Canada] was saying, *We took their ship away. They can't do campaigns anymore.* The ship is old. It's 50 years old. It's over. We've got the *Robert Hunter,* aka *Steve Irwin,* right now, and she's our flagship. She was called the *Westra,* a former Scottish fishery protection vessel. Then we called her the *Robert Hunter,* after the late … founder of Greenpeace. Then we renamed her the *Steve Irwin.* I should say "*Crocie,*" in honor of … the Australian crocodile hunter and environmentalist.

In 2007 Sea Shepherd founder Capt. Paul Watson received the Amazon Peace Prize for his work on behalf of the environment and marine species in Latin America. Watson was selected by Ecuador's Vice President Lenín Moreno and the Latin American Association for Human Rights. The award honors his dedication to protect marine wildlife around the world and his work in the Galápagos since 2000. Meanwhile, Sea Shepherd's campaign to stop illegal whaling in the Antarctic is heating up as the Japanese fleet fights back. This drama on the high seas is now being viewed by millions on the Animal Planet television series *Whale Wars.*

As for the shark, a creature that has traversed the oceans for more than 400 million years and that predates the dinosaurs, studies show that most species have declined by about 80 percent in the last 30 to 50 years, mostly because of human activities. "I often wonder if humanity is worth saving," says Cornelissen, who has worked so hard to save the sharks in the Galápagos. "On the other hand, when you see the beauty of the world I do believe we can still turn the tides if more people become aware of what's really happening."

CHAPTER 13

The Old Man and the Sea

"Many [fish] all their lives without knowing that it is not fish they are after."

—Henry David Thoreau

A small fishing skiff passes tour ships anchored in Wreck Bay and heads through mist toward a nearby dock. It's 5 a.m. and the sun has not yet risen over Puerto Baquerizo Moreno, the provincial capital of the Galápagos on San Cristóbal Island. I wait on the dock with four university students who are part of my environmental writing class at the nearby Galápagos Academic Institute for the Arts and Sciences (GAIAS), where I've had a teaching assignment for the summer of 2006. As the boat approaches an old fisherman rises from his seat and calls out, "Good morning. Welcome!" One by one we climb into his wooden boat, its green paint sloughing off like thin peels of cucumber.

Carlos Ricaurte is dressed in a white T-shirt, a blue windbreaker, and a tattered baseball cap. Deep lines radiate from the corners of his eyes from years of squinting in the sun. Looking down, I notice he's barefoot. We sit together on wooden seats as the boat heads out into the open sea. On the starboard side a baby shark swims next to a pelican that floats on the turquoise waves. A slight breeze comes up, and soon the sun rises like a lantern, casting a rosy glow over the old man's face.

Ricaurte is a subsistence fisherman who relies on what he can catch to feed his family. He's short and stout with three day's gray stubble on his chin. And while his face is creased by the sun and wind, there's a sparkle in his eyes, a liveliness born of the sea. Most fishermen in the Galápagos don't trust foreigners, but on this cool morning Ricaurte has agreed to take a small group of visitors out on his boat. Why? Because this 64-year-old has seen the future and it doesn't look promising. Most of the pelagic fish such as tuna and grouper have been fished out, and on many days there's no catch at all. Ricaurte admits that as a fisherman he has been part of the problem. Now he wants to become part of the solution, and he's created a new business here in the Galápagos: *pesca vivencial,* or lifestyle fishing. Twenty years ago, he says, fishermen could catch a thousand pounds of fish a day. Now they're lucky to catch a tenth of that, so he's begun taking tourists out to observe a dying way of life.

Today, his friend Carlos Vásconez has come along to help, and as Carlos maneuvers around piles of fishnets on the floor of the skiff, he seems younger and more agile than his 41 years. He's thin and tanned, and he recently buzzed off his hair to stay cool under the relentless sun. The men attach lures to four long rods and pull out about a hundred feet of line, which drifts behind us. Then Ricaurte removes two of the rods from their braces and hands them to my students. He's told us what to do if a fish bites, and we gaze at each other with a mixture of bemusement and anxiety. Some of these fish can weigh more than 80 pounds. Besides, fishing in the protected waters of the Galápagos Marine Reserve is strictly off-limits to tourists. The reserve, established in 1998, allows this right only to artisanal fishermen who live here, and authorities still haven't clarified the details of this new form of tourism. The Charles Darwin Foundation supports the tourist alternative but insists that regulations must be clarified and the activity monitored. Traditional fishing boats must be adapted to carry visitors; fishermen must learn to work with tourists and to speak English. It's a huge transition.

As we plow northeast of the island, Vásconez points to a basalt spire that, for obvious reasons, is called Five Finger Rock. "The locals call it Shit Rock because it's covered with guano," he adds, grinning. Suddenly, Ricaurte notices something moving just below the surface of the waves. The old man slows the engine and eases forward. *"Atún!"* he shouts. "Tuna!" "Mucho atún!"

"How can you tell?" I ask, squinting into the sun.

"Because of the way the water moves."

The swells roll like quicksilver as thousands of anchovies swarm in unison to ward off a school of tuna. The sea sounds like a boiling teapot. The fishing rods bob, yank, and sway, but the tuna aren't biting. Ricaurte is discouraged. "They're already full," he grumbles. "There are too many anchovies."

He and Vásconez pull in the lines and we head out to a world-famous diving site called León Dormido (Sleeping Lion), also known as Kicker Rock. The formation is the remnant of an ancient lava cone sliced in half by the sea: Two vertical rocks support frigate birds, blue-footed boobies, and red-billed tropicbirds. A small channel flows between the lava walls. On an outing like this, if the fish aren't biting, snorkeling is an option. We don our snorkeling gear and fall backward, one by one, off the boat into the chilly water. Almost immediately I see a spotted eagle ray hunkered down in the sand. Farther out a small green sea turtle flaps by, its carapace covered with algae. As we enter the channel between the vertical walls a strange new world comes into view; one populated with bright blue parrot fish, yellow-tailed surgeon-fish, and bizarre creatures called hieroglyphic hawkfish, which resemble something out of a child's coloring book. Green sea anemones cling to rocks. Sea stars brighter than navel oranges hide in crevices. Black sea urchins float like mini-Sputniks, their pencil-thin spines bobbing on the waves. At the end of the channel in this nautical Eden, a small Galápagos shark passes by but pays us no mind. We stare at each other through our masks, astonished, then climb into the boat and head back toward San Cristóbal.

Ricaurte refuses to be "skunked," so he attaches a lure to his rod and casts it into the sea, but something on the port side of the vessel catches his eye. He stands on the wooden seat and yells to his younger companion: "*Las palomas!*—The doves!" "Follow the doves!" He's referring to the black storm petrels that hover over the ocean like a storm cloud. The water bubbles as scores of anchovies try to evade the birds above and whatever's chasing them from below. Vásconez grips his rod and casts. His line immediately goes taut. It's a bite. He leans back and struggles to reel in his catch—a plump yellowfin tuna. Once he lands the fish, he releases the hook from its mouth and smacks its head against the gunwale. Blood runs red onto his hands and into the boat. Ricaurte nods with approval, throws him a plastic bag, and invites us to dinner at his house the next evening.

On the trip back I ask the fishermen about their lives. Ricaurte has nine children; all his sons are fishermen. Vásconez has three young children but does not want them following in his footsteps. "I want them to attend college and get an education," he says. To that end he's studying to become a dive master, but he's worried about passing the exams. But he knows he must. These days fishing is just too sketchy.

"What's the best time of day to fish?" I ask.

"When there's a quarter moon and it's dark. Light spooks them. It also depends on the time of day and the tides."

Suddenly Ricaurte's cell phone rings. The old man removes it from his pocket.

"*Ah-LO?*" he says.

Our eyes widen. Cell phones have caught on in the islands and almost everyone owns one, even fishermen out at sea. Ricaurte cuts the call short and rejoins the conversation. He's full of surprises, especially when he tells us that at the age of 18 he embraced the Baha'i faith, a religion founded by Baha' Allah in 19th-century Persia. It teaches the spiritual unity of all humans and holds that as a collective, humanity must seek peace and justice in the world. Ricaurte says he's one of about 30 Baha'is in the Galápagos.

"What attracted you to this religion?" I ask, intrigued by his belief system.

"Baha'i teaches that all faiths are equal. All races are equal. All men and women are equal. We're all connected to God. The world is a garden with many sweet-smelling flowers. It teaches us to preserve nature because we're part of it. We're all crying and suffering a lot, but if God is with us then death does not matter. We have to be conscious, to treat each other well because there are a lot of ugly things happening in this world."

I look up and notice a brown pelican following our stern. "He's waiting for us to fish so he can get some of the by-catch," Ricaurte says.

Carlos Ricaurte's house is a couple of blocks away from Wreck Bay. It has a corrugated sheet-metal roof, and its walls appear to be the same color green as his boat. I've come with my students from GAIAS, a ten-minute walk from Playa Mann, where sea lions laze next to sunbathers and where, if the water's clear enough, you can snorkel with sea turtles. Diego Quiroga, co-director of GAIAS, and his wife, Tania, are with us. As a cultural anthropologist Quiroga is drawn to this shift in identity among local fishermen. He and his colleagues have been working with Ricaurte to help him focus his objectives. Quiroga also set up our fishing excursion the day before. As we enter the fisherman's house the aroma of sizzling tuna fills the air. Wilma, one of Ricaurte's daughters, stands before a propane stove tending several pots. It's a small kitchen that opens to a living area with a tiled floor. The stucco walls are bare and patched in places, and the dining room table is covered with a flowered tablecloth made of polyester. Ricaurte's wife is in the hospital on the mainland, recovering from an amputated toe caused by diabetes. His friend Vásconez has arrived, but he has gone down the street to buy a dozen tall Pilsners. When he returns, Ricaurte comes to the table with a platter of steaming yellowfin tuna and serves each of us a portion. Wilma then piles our plates high with rice, shredded carrots, red onions, and wedges of fresh lime from the highlands. Ricaurte takes his seat at the head of the table and raises his glass.

"*Salud!*" he says.

"Salud!" we chant in unison.

After dinner Ricaurte inserts a video about Galápagos wildlife into his VCR. He wants us to understand this place he calls home and why it's so special. Two of his grandsons sit on the tile floor. When an image of a red-bellied creole fish appears on the screen, one of them shouts, "*Un pez gringo!*" A gringo fish, named for all those sunburned tourists.

Ricaurte tells us about his plan to buy new boats for this new industry he's created: ten catamarans big enough to carry ten passengers each. But he needs a loan—a big one—and the proposition isn't easy. There's also the permiting process through the Galápagos National Park, which for many residents is a continuing nightmare.

In March 2008, about midway through my year of living in the Galápagos, I take a speedboat from my home in Puerto Ayora to Puerto Baquerizo Moreno: I want to meet again with Ricaurte and Vásconez and learn of their progress.

I check into the Hotel Mar Azul, walk a few blocks to Ricaurte's house, and knock on the door. It takes a moment before he recognizes me, then he invites me in. On this bright morning he's wearing a T-shirt with the logo, "Hard Rock Café, Sydney." He's still doing artisanal fishing, but he wants to expand his business to include scuba diving with a dive master.

"What about your friend Carlos Vásconez? Did he ever get his dive master's license?"

"Sí," he replies. "Right now he's out on a boat working as a dive guide."

Ricaurte is preparing to travel to the Galápagos National Park headquarters on Santa Cruz Island. He has a couple of dozen applications for boat permits from locals, including his own. He wants a permit that will allow his passengers to fish—and to keep their catch.

"Isn't that sportfishing?" I ask.

"No. I'm against sportfishing," he replies. "The millionaires have tried to cash in on this activity, but the local government stopped them. Sailfish should not be caught in the Galápagos, even if they're released. It's too much of a struggle for the fish. It becomes exhausted and weak, and if it bleeds it could attract sharks."

Ricuarte's main mission is to teach tourists about the natural resources, where certain species live and how they adapt. "But some things must be kept secret."

"Like what?"

"If I know where a lobster cave exists I'm not going to share that information with my passengers or anyone else. Fishermen do not pass out that kind of information to potential competitors."

Ricaurte charges $125 a day per passenger. He gives students a break at $75 apiece. Some days he takes out two or three passengers, and some days no one. He's frustrated that the park now requires safety classes and has stalled on handing out boat permits. "This is unjust. They've changed the rules of the game at the last moment to prevent us from going ahead."

Now that tourism is rising in this little town of 5,600 people, he finally has a chance to make a living. But all those vague regulations and the safety classes have him flustered. "Of course we need to have safety gear like life jackets and radios and comfortable accommodations. We can't put tourists on some old fishing boat. There's a new class of tourist coming to the Galápagos now."

Ricaurte staunchly believes that fishermen must change their livelihoods, but he says that he'll continue to fish on a subsistence level. "I'm not really changing my activity; taking passengers aboard is an alternative to extracting the marine resources. I started to see the resources dwindling and I worried about what would be left for my kids. If they have no other options, what will they do?"

In an adjoining room one of his grandsons blares out some notes on a dented trombone.

"Who's doing that?" he shouts through the wall. "Stop that noise right now!"

Ricaurte picks up his stack of applications for boat permits and waves it in the air. He's tired of being stuck in limbo, just like Jack Nelson and Mathias Espinosa at Scuba Iguana. "We can't continue like this," he says, his voice rising. "If I have to, to get a permit I'll go all the way to the General Assembly in Quito to make my case!"

In June I again return to Puerto Baquerizo Moreno to meet up with a new group of students and, I hope, to find Carlos Vásconez. To my surprise I run into him on the street. We walk to GAIAS where we sit outside on a rock wall overlooking the turquoise bay. He's so excited about passing his examinations that he pulls out his dive master's license and shows it to me. I extend my hand. "Congratulations," I say. "Very nicely done." Vásconez beams. His daughter has gone to the mainland to attend the Polytechnic Institute in Guayaquil. She wants to become a marine biologist, and one of his sons wants to become an architect.

"And your youngest son?"

"He wants to become a fisherman, but he's too smart for that."

A few days later we meet in a classroom inside GAIAS. Diego Quiroga joins us. Vásconez speaks freely of his life as a *pepinero*—a sea cucumber fisherman—during the "gold rush" days of the 1990s, when men were making so much money they didn't know what to do with it. Vásconez says he made "millions and millions of sucres." (This was before Ecuador changed its currency to the U.S. dollar.) In 1996 fishermen were lighting cigarettes with sucre notes worth about $50 or throwing money into the streets. Most pepineros blew their earnings on alcohol, drugs, and prostitutes. Santa Cruz Island is full of them.

"During the boom we would go to the *chongos* (whorehouses) and would wipe the tables with the bills, then throw the money on the floor and all the girls wound dive for the bills," he says. The town became known as Puerto Puta (Port Whore). "There would be a line of dinghies waiting for the prostitutes."

"Did you ever fin sharks after it became illegal in 1986?"

"Yes, when I needed to survive."

"How much money could you make?"

"In 2005, about $30 a shark." Today shark finners earn about $200 per shark for the dorsal and pectoral fins.

My next question is difficult. "How did you feel killing sharks?"

Vásconez shifts in his chair. Before he took scuba diving classes he justified shark finning because he was bringing money home to his family. But during a dive at Wolf and Darwin Islands he was surrounded by hammerheads. "When I looked into the eye of a shark and realized how special it was, I felt I had committed the worst crime in history. I was very ashamed. Now I can really relate to the species and I understand the need to protect it." He says it's important that other fishermen understand this and change their livelihoods by becoming dive masters.

I admire his candor. "Carlos," I ask. "Have you ever killed sea lions?"

"Yes, we would cut a sea lion into pieces and tie it into a net to attract sharks. Now, fishermen use more mullet than sea lions."

Despite his rise in status as a dive master, Vásconez says it's very difficult for a fisherman to even obtain a boat. Permits are based on money and a complicated points system. "If you don't have a boat or a scuba diving center you make very little compared to the guy who's managing the center. The park gave us this opportunity to change our livelihoods, but in reality it hasn't helped much. They train us but at the same time they deny us access to a job because of all the restrictions."

And while tourism grows all around him, he laments the fact that corrupt fishermen and the big tour companies are profiting. "There is only hope for those who have money and boats, but not for the rest of us. The conservation sector in the park administration should not stop us from working. Otherwise all they'll achieve is to create big cities with a lot of criminals and thieves."

CHAPTER 14

Birdland

"And a good south wind sprung up behind;
The Albatross did follow,
And every day, for food or play,
Came to the mariner's hollo!"

—Samuel Taylor Coleridge

It's 5:30 a.m. in May 2008, and Kate Huyvaert has just arrived on the dock in Puerto Ayora with a truckload of metal boxes and camping gear. She and a group of biologists are headed to Española Island in the southeastern part of the archipelago. Huyvaert is a pretty young woman with a natural smile and auburn hair pulled back in a ponytail. She's a conservation biologist from Colorado State University and has come here to monitor the largest colonies of waved albatross in the world.

This elegant species—a perfect flying machine—spends most of its life in flight and can glide for hours without ever flapping its wings. The waved albatross is endemic to the Galápagos, where it breeds on Española Island. It's now critically endangered, and illegal longline fishing in the archipelago and elsewhere has killed more of the birds than Huyvaert likes to imagine. One of her missions for the next two weeks is to retrieve special electronic units called geolocators that were banded to the legs of three albatross about six months earlier. The data she collects will show, based on light angles and water temperatures, exactly where and how far the birds have migrated while feeding out at sea.

Our speedboat heads out past dozens of tour boats anchored in the harbor. On board with Huyvaert is her biologist husband, Paul F.

Hoherty, Jr., and two other albatross scientists from Syracuse University who will explore the island's rocky interior. Wildlife photographer Tui de Roy and her partner, Alan, have also come. The sea is rough and choppy today, and two of the researchers hang their heads over the gunwales for most of the trip. This is not good: It's a four-hour trip to the island and the currents are wicked. At mid-morning we arrive at a tiny beach called Manzanillo, where we drop off the team from Syracuse University. As our boat heads back out, we can see them standing there frozen in place, two tiny specks staring into a tangle of brush. To reach an albatross colony in the hot and waterless interior of the island, they must hike through a fortress of brush.

Our own destination is Punta Cevallo on the southeastern edge of Española, a coastline so rocky it seems impossible to land. But the captain knows this beach well. He's brought Huyvaert here before, and as he ferries us to shore, he praises her as the *madrina* [godmother] of the albatross. "She looks after them," he says, smiling. "She's their guardian." The beach is short and narrow, more so because of the wooden boat that sank here some years ago and now lies scattered in pieces. Sea lions love sleeping on its water-polished hull, away from the bony coral that litters the sand.

We make camp just beyond the wreck, but we're not alone. A large nesting albatross sits on its single egg just 20 feet beyond our "kitchen." Remarkably, the bird is unfazed by our intrusion, its eyes closed in slumber. Hermit crabs scuttle by, almost too fat for their shells. The mockingbirds are famous on this island for pecking at water bottles. They perch on our tents, daypacks, and chairs, peck at our sandaled feet, and try to drink from our cups. If a water pitcher is left uncovered, Huyvaert tells us, "a mocker can fall in and drown." An unused cup is always turned upside down.

Punta Cevallo rises from the surf just beyond camp, a promontory above an aquamarine sea where waves crash against basalt. It's also home to the largest known colony of waved albatross in the world, a sanctuary where pairs reunite, perform dazzling courtship rituals, and share the arduous task of incubating their egg in the blazing sun. When the female needs to leave the nest to fish, the male takes over the duty. When she comes back, it's her turn to tuck the egg beneath her downy feathers. And so it goes. It's one of nature's loveliest spectacles, one that few people ever witness. The colony is closed to tourism. In the morning I'll follow Huyvaert through the colony to watch her band the newcomers, which is no easy task. The birds are the size of small turkeys, weighing about 7.5 pounds with a broad wingspan of nearly eight feet. About 3,000 breeders inhabit this island. All must be monitored, but only three have been fitted with geolocators; the units, which give these researchers the time of day and the exact location of the birds to determine where the albatross feeds and when, impose little discomfort on the birds.

Daybreak. Huyvaert has already finished breakfast and has packed her field gear for a day of work. We follow a sandy trail up to the promontory. In the pink light of morning, albatross appear everywhere. I follow the young biologist through this world-famous rookery, trying not to get too close to the birds as she checks the tags on their pale-blue legs. At the first nest, the albatross clicks its beak in warning. It sounds just like the clacking of two sticks, a musical instrument from the tropics. First she uses a pole to move aside one of the bird's wings and see if it's been tagged. If not, Huyvaert puts on a blue gardening glove and grabs the bird by its bright-yellow beak—quickly, to avoid being bitten. Then she wraps an elastic hair tie around its beak, lifts the heavy bird into the crook of her arm, and tags its leg. The albatross struggles. *Ha-honk!* it utters, dropping guano down her khaki-colored shorts. "I know, I know," she tells the bird in soothing tones. When she sets the newly tagged albatross down, it settles back on its egg as though nothing has happened. "Good job," she says.

I ask her if she's ever been bitten. "Oh, yeah," she responds. "It's like a really bad paper cut. They bite you and you jerk your hand back. They've got real sharp-edged bills. But if you let them clamp down and you lift off their jaws, it's not so bad."

Next she grabs a squirmy male. Shocked to be off terra firma, he frantically waves his webbed feet in the air. "Easy. Settle down," she tells him. He's incubating the egg while his mate is out foraging for squid and fish for the next couple of weeks. As Huyvaert tags the bird, thin black lice crawl from its feathers down her shirt. "It's not very glamorous," she says of her work, grinning and brushing them off.

Huyvaert moves to the edge of the cliff and sits on a rock to record data in her logbook. A strong breeze blows in from the south, ruffling the hair she's tucked through the hole at the back of her cap. I sit next to a pair of albatross engaged in a mating ritual so elegant I'm mesmerized. First they bow to each other like Buddhist monks, their saffron beaks pointing downward. Then their bills trace circles around each other. They clack back and forth in a language of their own and begin to dance. One foot up, the other foot down. At times they point their beaks skyward and utter a call that sounds like *mu-hoo.* Their feet are bluish, their heads gray with tinges of gold. Wings and tails are chocolate brown, and the feathers beneath their masterful wings bear elegant waved patterns, hence their name.

The albatross spends most of its life out at sea. It returns to this rookery only to breed and raise its chicks until they're ready to fledge. When Huyvaert was working on her Ph.D., she monitored the nests and eggs every single day for about two months. "One day you come along and there's this little hole in the egg and there's this little thing sticking out and moving. Then the next day there's a much bigger hole and it's halfway hatched out."

Once a chick has emerged, the parents brood it in a special patch on their bellies to keep it warm. Early on, they feed it tiny amounts of oil. I ask Huyvaert where the oil comes from. "This is a really cool thing about the physiology of albatross. Part of their digestive tract is called the proventriculus [a gland in the stomach], which acts like a separatory funnel. They'll go out to sea and eat something—squid, fish—and the proventriculus separates all the gook to make oil." So, instead of the parent being full of fish when it returns from sea, it carries that oil, which it regurgitates to the chick.

The chick is brooded for up to two weeks until it's about the size of a small football. For six months, the chicks in the colony hide out in the shade while their parents take turns foraging off the coast of Peru, returning every few days, then weeks. The young birds spend a lot of time in the colony, flapping their fuzzy white wings, testing them by hopping from rock to rock in little spurts, gaining a bit of uplift before coming back down. "I've seen them doing that for days," Huyvaert says. I've also seen this phenomenon. A few years ago at Punta Suarez on the northwest coast of the island, I sat with my tour group on a cliff near a huge blowhole, watching albatross use the bluff as a runway. They'd race toward the edge of the cliff on gawky legs, their wings teetering on takeoff, then float off effortlessly into the sky.

The promontory at Punta Cevallo, on the other hand, is birdland at its best. It teems with life today, and not just "albies," as Huyvaert calls them. Blue-footed boobies whistle and point their beaks skyward while stomping their rubbery feet. It's as goofy as something out of Dr. Seuss. When the female accepts the male, he picks up a piece of straw and lays it down before her. This simple offering seals the deal: It's the beginning of their nest, where the pair will share the duties of raising a chick.

On the other side of the bluff, Tui de Roy and Alan are exploring—on a sort of camera safari. One of de Roy's goals is to capture birds in flight close-up. It's not easy. The light can be flat. A frigate bird can fly right out of the frame. A red-billed tropicbird can turn its head, obscuring the very feature that gives it its name.

That night, as we sit around the plastic dinner table, we snuff out the lantern. The sky burns with a million stars. The Milky Way galaxy is a brushstroke in white. De Roy seems disappointed in her day's work, first because the light was flat and uninspiring. Plus, after a lifetime of film photography, she has just switched to the world of digital and is still getting the knack of the "new" technology. I ask if she ever does night photography.

"I've tried, but I've never had much success," she says. Then she astounds all of us by adding, "I don't really know that much about photography at all. I know wildlife and how it behaves, but I honestly don't know the technical aspects of photography."

"But you understand light and composition," I say.

"Yes," she replies, her eyebrows furrowing.

Clearly, de Roy's lifetime experience has little to do with F-stops or the Ansel Adams zone system. It's about waiting for that decisive moment and knowing exactly when it arrives. Even so, she's not happy with the day's images. Digital, she contends, can make it too easy to succumb to grab shots. "With film there are no wasted shots." She learned this long ago as a child growing up on Santa Cruz Island. And so every evening she trudges back to camp two shades darker with her ten-gigabyte flash card full of images, which she downloads to her laptop.

The next day something clicks. The images she's captured are some of the most stunning I've ever seen: swallow-tailed gulls in a rocky cove, their red-ringed eyes ablaze; close-ups of a red-billed tropicbird from every angle in perfect light; a frigate bird gazing straight down into the lens.

Back in my small tent, I have the sense that I am poised on the outer edge of the planet. I drift into new realms of imagination. Española's wildlife is the most diverse and vibrant of any island in the Galápagos; this is the oldest island, and on a path to destruction. Eventually, it will sink beneath the sea and move eastward toward the South American continent, just as the rest of the Galápagos Islands will. In time, others will rise up from the Pacific floor to take their place. Through the mesh of my tent, under my single bed sheet, I watch a web of faraway stars whose twinkles burned out too long ago to fathom but are still reaching Earth. In the illusory worlds of time and space, I am but a speck of dust astride a dying rock out in the wild Pacific. I nod off to the sound of surf, to night herons crunching on crabs, and to the plaintive cry of a young sea lion bleating for its mother like a newborn lamb.

The mockingbirds here are relentless. I wake to a pair that have perched on my tent frame, squawking as the sun pushes clouds. This subspecies of mockingbird is slightly larger than the other three species in the archipelago. It has long legs, a curved bill, and reddish eyes. The birds tend to gather in groups but are reluctant to fly, preferring instead to hop about in little spurts. I rise and walk the short distance to the camp kitchen past saltbush and other coastal plants. The nesting albie that first greeted us is still there, sitting on its egg. I drink a quick cup of coffee, slather on sunscreen, and return to the bird symphony up on the bluff. Swallow-tailed gulls, the only nocturnal-feeding gull in the world, scream with mouths wide open, their ruby tongues clearly visible. The gull's head is black, its eyes encircled in red. While feeding at night, it emits a kind of clicking sound that some researchers believe may be a kind of echolocation to help it find food.

Masked boobies squawk from the edge of the cliff. These are the largest of the booby species in the Galápagos. The mothers lay two eggs; both hatch, but only one chick survives. The older one pushes the younger one away to starve, or kills it outright. Scientists call this "siblicide." According to David J. Anderson, a biologist at Wake Forest University who discovered this peculiar behavior, it is key to the species survival. Masked booby parents, he found, actually encourage the older chick to kill its younger sibling. Why? The short-term sacrifice means fewer off-spring but a healthier population, and that ensures survival. This is how it works: The second egg usually hatches about a day after the first. When the younger sibling breaks through the egg, the elder chick nudges it out of the nest to die of dehydration in the sun, or be eaten alive by hungry mockingbirds. The firstborn may kill its sibling outright. So why lay two eggs? "Insurance," says Anderson. If the first egg does not hatch, the second provides fall back. The effort needed to raise two chicks in this harsh environment is just too great. A second chick is a liability.

Anderson and his students have found out why. "Parents in this species have crappy hatching success," he tells me. "Why? We don't know. To assure that even a single chick survives, they have two as backup. This gets them to 85 percent probability that at least one will hatch. Sometimes two hatch, but if they try to raise two, moms have elevated mortality and the chicks are scrawny, performing poorly as adults. So, hatching two chicks constitutes a mistake. Siblicide corrects the mistake."

Nazca boobies also have a rather crazy mating system, he adds. "When a mated pair raises a chick successfully, the female shows the male the door and mates with another male. You see an inverse correlation between a male's success and his immediate future prospects to mate. Of course, that also means that anyone currently out of the mating pool is likely to get back in shortly. You might have expected that successful males are valued by females, and that is the case in most mating systems. In this case, the sex ratio is biased, with an excess of males, and successful males incur reproductive costs that current non-breeders do not incur. Females jump on a fresh horse by taking up with a male who hasn't incurred reproductive costs recently."

Nazcas have a lot of extra-pair copulations, a bit like the albatross, says Anderson. "Before egg-laying, a female will be standing with her betrothed and then walk to the next nest site and copulate with the male there, then walk back to her betrothed. Not much he can do about it because he's smaller than she is. Until eight days before the first egg is laid, she's getting most of her action from neighbors, and it's a *lot* of action. But then eight days before, she switches to just her main squeeze of that year, and all of the babies are his. The extracurricular sex is probably maintaining relationships with other males ... to whom she will switch in the future when she has gotten what she can out of the current male."

———————————

One day near the edge of the bluff, Huyvaert finds an albatross with a steel tag on its leg that reads: da@wfu.edu. That's the ID of the same Professor Anderson—the man who has come to this very camp for decades to monitor the albatross and masked booby. The tags bear his e-mail address so that he can easily be contacted by someone who finds one of these birds, dead or alive. Anderson is one of Huyvaert's mentors. In fact, if not for his influence, she might have become a medical doctor, like her grandfather before her. As an undergraduate student at Wake Forest she was majoring in both English and biology. When a snowstorm marooned her during Christmas break, she couldn't register for a class she needed. Huyvaert called Anderson, who at the time, was her biology adviser. He convinced her to take his evening ornithology course. "It was like, 'Birds?' " she recalls. "I'm going to study *birds?*" The class changed her life and provided critical data on why the albatross is now endangered. While the bird is protected by the Galápagos National Park, its range is limited and its numbers are threatened by the impacts of tourism, disease, climate change, and especially illegal longline fishing. One day on the island, Huyvaert and her husband found an albatross with an enormous hook lodged in its throat. It had come from an illegal longline. The two carefully removed the hook, but not before photographing it as evidence of human impact on the species.

Huyvaert and I sit at the camp table after she's logged the day's data into her laptop. I ask what inspired her to study nature, whether there was some defining moment that stirred her. It was more like a pattern of events, she says, tweaked unexpectedly by fate. Huyvaert grew up in Cincinnati where she spent long days at her grandparents' house, mucking around in the creek that ran through their property, mostly looking for fossils. On holidays she'd arrive at her aunt and uncle's house wearing a cute little dress. "I would wear it like for five minutes, change into play-clothes, and then go outside with my boy cousins and play."

"You were a tomboy?"

"Oh, yeah."

Huyvaert first visited Española in the summer of 1994 as an honors student and field assistant for one of Anderson's graduate students. The island and its albatross, a species believed to mate for life, had intrigued her and she set out to help count every albatross on Española. What the researchers learned about this species turned previous assumptions on their heads. Sometimes Huyvaert and her field partners would see individual albatross chasing others, knocking them down, and attempting to mate. "You'd see three and four car pileups in this fray," she says. "So you wonder, 'Well, what is *this*? What's going on?' "

Contrary to the long and dearly held belief in avian monogamy, scientists found that one-quarter of waved albatross in the Galápagos actually get, well, "divorced." In 1995 Huyvaert and her partners took blood samples from 30 albatross families and sent them to the United States for DNA "fingerprinting." Anderson credits Huyvaert for discovering that 25 percent of the chicks they sampled weren't related to their social fathers. But Huyvaert says "it was a group effort" between Anderson, an ornithologist named Patricia Parker, and herself. The discovery was so fascinating that it turned into a dissertation for the budding biologist. Working with her adviser, Dr.

Parker, Huyvaert embarked on a three-season study of every albatross parent and chick on Española Island. One reproductive theory is that females have extra copulations to avoid inbreeding with a male that's too closely related. A female might have a mate, but it's likely she'll leave the nest at times to find the genes for her offspring from a different male. "It's pretty complicated," Huyvaert admits. "There are a lot of unanswered questions."

Some of the questions will indeed be answered once the geolocator data are analyzed. Her husband, Paul Hoherty, recovered the first geolocator, and I am with Huyvaert when she finds the second. As always, she secures the bird's beak with a hair band and scoops the bird up in her left arm. Using pliers, she releases the data-bearing electronic unit and sets the albatross down. "Hey, buddy," she exclaims, obviously excited, "Thanks for your service to science!" The albatross seems unfazed and goes right back to sitting on his egg. By the end of the day, Huyvaert and her husband will have located about 400 albatross. A few days later they discover the third geolocator, but the data are still being analyzed.

The waved albatross may seem like a brawny bird, but it's more vulnerable than one might think. A 2001 survey revealed that about 60 percent of this population tested positive for adenovirus, something like a common cold. That's not a big deal, but if an El Niño event were to hit, it could be disastrous. If too much rain falls, the vegetation goes wild and the albatross is likely to pass the island right by. It can't nest under such conditions; it needs open ground. Second, too much rain attracts mosquitoes, which harass the birds and force them out to sea where they abandon the effort to breed. Third, warmer water temperatures can deplete the marine life the albatross depends on for survival.

On the last day of my visit I walk out to the edge of the cliff and gaze out over the ocean. White waves slap against stones the color of chocolate. In some places, the lava is so gas-blasted that it resembles gargoyles. Weird faces appear everywhere in the 3.4-million-year-old rocks, while down in the battering surf, marine iguanas with bellies full of algae battle the tide to gain a foothold in this seemingly lost world.

At noon, de Roy, Alan, and I must return to Santa Cruz Island. The speedboat arrives right on time with the two researchers from Syracuse University who've been working inland, another researcher from Quito, and our old hiking pal and photographer from Volcán Alcedo, Pete Oxford. Huyvaert and her husband will stay on for another week before returning to their country farm in Bellvue, Colorado. We bid them farewell and plow through waves, avoiding the deadly rocks that likely caused the wreck on the beach where sea lions sleep.

Back in Puerto Ayora I meet up with de Roy and Alan at Angermeyer Restaurant, an upscale establishment on the bay owned by one of the Galápagos' earliest families. De Roy's newest book, *Albatross: Their World and Ways,* co-authored with Mark Jones, her longtime business partner, has just been published, and that's cause for celebration. I ask what it is about the albatross that attracts her. After all, she and Jones traveled the world's oceans in a 43-foot sailboat to document every species of albatross, surviving storms and a near shipwreck near Antarctica.

De Roy describes her love for the albatross like this: "I'm really taken by all things that are beyond humanity. To me it's as close to heaven as I can get—and I mean 'heaven' metaphorically. True wilderness to me represents perfection, and the biggest problem that albatross have is the fact that we humans actually never *see* them in their rightful element. We see them sitting on a rock, on a beach, on an island, which is their least favorite place. They do it because they have to, to raise their families. But that's not where an albatross belongs. An albatross belongs in midair over the ocean. I can see it, but I can't reach it. It's heaven, but it's just there at my fingertips. Here is a bird that lives an amazing life and that is incredibly beautiful and full of dignity and grace and mystery. But I can't get close to its realm."

It's growing dark as we sip our drinks, and tiny geckos emerge near a string of lights on a post next to our table. I ask de Roy whether she believes the Galápagos will survive as the so-called last paradise.

"Galápagos hasn't suffered a biological holocaust yet, and because it hasn't happened yet we are lulled into thinking that it's not going to happen. It could happen tomorrow. It could be happening right now as we sit here. The dengue mosquito, the West Nile virus mosquito, whatever could be doing its dastardly deed right now as we sit here. When it really starts to happen, it's gong to happen fast. There are all these biological time bombs that are sitting there ticking, except we don't know when the clock was set."

She goes on to remind us that biological holocaust doesn't measure things in moderation. Extinction isn't a gradient; it's a reality.

"It boils down to the fact that the isolation bubble has burst. The islands originally were close enough to the continent to receive a lot of species, and far enough for those species to lose contact with their origins, which is why we've got this phenomenal biodiversity. The isolation is no longer there. We have popped that bubble."

How? By creating a direct link with the continent, where biological agents can travel on the endless daily flights arriving in the Galápagos, on cargo ships, on the soles of people's shoes, and in the water or air. One tiny vector, that's all it would take.

Even so, de Roy is cautiously optimistic. "The trump card is that nature is actually more resilient than we think and that's a trump card we don't want to play too soon. We don't want to say, 'It's not as bad as it sounds,' because it *is* bad. Very often nature gives you a second chance, and second chances give you call for optimism. So, no, I don't think Galápagos is going to vanish. I think we're coming mighty close to testing that theory, because by the time we find that we've lost it all, we won't have a second chance. It isn't black and white. But I think there's hope, and hope is what we have to cling to."

CHAPTER 15

God on the Rocks

*"Theology made no provision for evolution. The biblical authors
had missed the most important revelation of all! Could it be they
were not really privy to the thoughts of God?"*

—E. O. Wilson

A few years ago I took a day trip to the fantasy island called Bartolomé, whose volcanic landscape resembles Mars. The island is one
of the rawest examples of geology in the Galápagos, a tapestry of red
cinder cones, volcanic bombs, lava flows, and tunnels running right
beneath our feet. My naturalist guide that day was a man named Victor
Rizzo who spoke rudimentary English and sidestepped questions about
how life had evolved on the islands. The tour group included visitors
with higher degrees, including one in science. Perhaps it was Victor's
level of fluency but it seemed we'd hit a nerve—the "E" word. Later I
took him aside and spoke to him in his own language. He admitted
he did not believe in evolution. He *did* believe in adaptation but not
the descent of our own species from a single ancestor. He said he was
now studying the bible of the *testigos,* the Jehovah's Witnesses in Puerto
Ayora. Victor was in fact a creationist, a biblical literalist, who believes
Earth is only 6,000 years old and that humans were created in the same
bodies we now inhabit. Science writer Natalie Angier echoed this belief
in her book *Canon: A Whirligig Tour of the Beautiful Basics of Science,*
when she wrote that creationists believe "God placed humans here as
is—*in toto*—and that goes for Dorothy and her little dog, too."

This was one of the biggest paradoxes I'd ever heard in Charles Darwin's cradle of evolution, and I soon began looking into the phenomenon. What I learned was astounding, not only in the Galápagos but in my own country as well. A Gallup Poll conducted in 1997 revealed that a majority of Americans are creationists: Forty-four percent believe that God created humans in their present form within the last 10,000 years. Thirty-nine percent believe in "theistic evolution" or "intelligent design," meaning that God or some higher form of intelligence played a role in our creation. (Most scientists call this a subtle cover-up for fundamentalist creationism.) Only 10 percent believe in "naturalist evolution," the scientific view that humans evolved over millions of years from single-cell organisms, and that God had no hand in it whatsoever.

A recent Gallup Poll from 2008 shows those numbers have not changed in the last decade. In high schools throughout the nation creationism is still widely taught as a sound alternative to evolution, according to a poll published in the journal *Public Library of Science Biology*. And that seventh wonder of the world, the Grand Canyon, stirred up a scandal when the national park kowtowed to pressure and began selling a creationist book denouncing evolution. The Institute for Creation Research takes its own tours through the Grand Canyon and teaches that the chasm is only about 4,500 years old—and that it was formed by Noah's Flood. Most scientists believe the canyon is six million years old. As for "intelligent design," a majority of scientists call it "pseudoscience." Others, such as the American Association for the Advancement of Science, deem the belief "junk-science" and "devolution."

When I returned to the Galápagos I looked for the guide from Bartolomé Island but learned he was on the mainland, so I found other testigos and even attended a "meeting" with them at Kingdom Hall in

Puerto Ayora. I wanted to learn about their beliefs and lifestyles, especially among those who worked as naturalist guides for the Galápagos National Park. How did they resolve the discrepancies between faith and science on islands almost synonymous with evolution?

One morning while I'm talking with Jack Nelson on the pier outside the Hotel Galápagos, a longtime resident named Marco Antonio "Tony" Aguirre stops by to visit. Nelson is busy adjusting the new "green" engines he's bought for one of his diving boats for Scuba Iguana. He and Aguirre have known each other for decades, and while they respect each other, their perceptions of life on Earth span a rift as wide as the galaxy. Aguirre is a highly esteemed Jehovah's Witness in Puerto Ayora. He's a diminutive man who has ridden his bicycle down here from El Peregrino, a tidy bed-and-breakfast he runs with his wife, Kathy. Aguirre dismounts and stands in the shade of an overhang while Nelson turns wrenches inside his new engine. When Nelson is finished he wipes his hands on a cloth and walks down the pier to greet Aguirre. I follow.

"Hi Tony. What's up?" he asks.

"Oh, not much. I just stopped by to say hello."

Aguirre extends his hand and greets me in perfect English. His words are soft, his inflection kind. It turns out that both our fathers were pilots. Aguirre's father was Ecuadorian but died when the mail plane he was flying crashed into Chimborazo Volcano in the Andes. His mother came from Nebraska, where he lived for awhile after the accident. Aguirre has lived in the Galápagos now for about 35 years. He asks me what it was like growing up in foreign countries like Japan and Spain. He seems as curious about me as I am about him, and we make plans to visit at his B&B. I want to learn more about his life as a Jehovah's Witness and why his faith denounces evolution. I also want to meet his son Joed, a naturalist guide for the Galápagos National Park. Aguirre is happy to oblige.

On a cloudy morning I walk up a flower-lined path and meet Aguirre at his El Peregrino. He's sitting at a table on the patio, drinking tea and studying Quichua, an indigenous language from the Andes. I ask what his interest is in this difficult tongue. "It's a lot easier to reach them," he says, referring to some Otavaleño and Salasaca Indians who have recently moved to the Galápagos. "They tend to shy away from Spanish speakers, so we find that if we address them in their own language they're more receptive."

One of the requirements of being a Jehovah's Witness is door-to-door "ministry" to educate those in the community about Jehovah and to distribute literature such as *The Watchtower* and *Awake!* Witnesses are conservative fundamentalists whose religion is an offshoot of Christianity. They interpret the Bible literally, including Eve's creation through Adam's rib, Satan's temptation in the Garden of Eden, and the great Flood that launched Noah's ark. In Genesis, they identify the world's creation in "eras" instead of seven 24-hour "days." Evolution, they claim, is just a myth. Congregations meet three times a week and members address each other as "brother" and "sister." Witnesses are allowed to drink alcohol (though not to excess), but are forbidden to celebrate birthdays and holidays because nowhere does the Bible encourage it. (I later learn how this belief system has ripped some families apart.)

Jehovah is the name of God, and Aguirre claims that it appears over 7,000 times in the Bible. It means "he who causes to be." It also implies purpose. "As the Bible explains it, the heavens belong to Jehovah and the Earth he has given to mankind. He told Adam that mankind was to take care of the Earth, to produce children and fill the Earth, not overfill it but to fill it comfortably. To take care of the Earth and make it a paradise." But mankind is destroying the Earth, he warns, and eventually God will intervene and destroy the destroyers. "The real danger is coming from Jehovah. He's very much against the way that man has mismanaged his inheritance. It's basically a matter of survival. That's why we go house to house—to get our personal lives in line with God's standards."

Aguirre shifts his baseball cap and smiles, his eyes shining through square-rimmed glasses. He doesn't believe in macroevolution, common ancestry through time, and this often presents problems when Witnesses work at the Charles Darwin Research Station. "Every once in awhile we come upon people in the scientific community who subscribe to Darwin's theory of evolution. They are very reluctant to get into any real discussion about it; they see it as a kind of polemic. But we don't find much of a problem here among the common people."

I mention the creationist guide I had on Bartolomé Island and how surprised the group and I were by his belief. "My son Joed is a guide, and he's a Witness, and he finds that problem quite often. Usually he doesn't explain the things around him in evolutionary terms."

I ask to speak with Joed and to attend a "meeting" at the Kingdom Hall with his family. "Yes, of course," says Aguirre. "You can sit next to my wife, Kathy."

The Witness Hall in Puerto Ayora stands on a backstreet a few blocks away from El Peregrino. It's a plain building whose interior resembles a classroom more than a church. The walls are completely bare. There are no icons, no crosses, candles, statues, pipe organs, or stained-glass windows. The only item at the front of the room is a plain wooden podium. As I enter the building, people welcome me and shake my hand. I take a seat next to Aguirre's wife, Kathy, in one of the plastic patio chairs lined neatly in rows. Men enter wearing suits and ties and women in ankle-length dresses. Soon a gray-haired man dressed impeccably in a gray suit, a lighter gray shirt, and a burgundy tie walks up to the podium. He's what's called an "elder," a highly respected leader in the Jehovah's Witness community. He's also Aguirre's nephew. The elder begins the meeting by saying that certain birds migrate thousands of miles without guidance and return to the precise places they once came from. "How do they know to follow the exact same routes? Who guides them? Scientists say they know the exact reasons for this intelligence, but we reject the scientific explanations. The one and only guide that directs them is Jehovah," he declares.

We're now instructed to open the Bible of the Jehovah's Witnesses and read a passage aloud. Kathy, a graceful woman with spiked gray hair, points to the verse in my book and I try to follow along. The Witness Hall is hot, and I'm feeling sticky, as brothers and sisters swat at mosquitoes with copies of *The Watchtower*. A little girl in a frilly white dress leaps from her chair, runs down the aisle, and is pursued by her sheepish father. Boys wearing polyester suits and ties squirm uncomfortably in their dress-up clothes, and one kid plays with strands of his mother's hair until she shoos him away. She's wearing shiny gold sandals with heels. I'm wearing an ankle-length dress and a shawl to cover my bare shoulders. It keeps the mosquitoes at bay.

We sing a hymn then open a study guide called *La Atalaya*. The elder at the podium reads passages from an article: "El Reino de Dios Pronto Nos Liberará" ("The Kingdom of God Will Soon Liberate Us"). Each paragraph contains a related question, and as the man asks for answers, hands shoot up like arrows. Sisters quote verbatim from the text, and brothers read it aloud. The elder comes to question 13: "How will Jehovah liberate those of his servants who are already dead?" The little girl in the frilly dress raises her hand. Aguirre's son Joed, who is working the microphone, walks up the aisle and holds it close to her lips. From where she sits on her mother's lap, the toddler answers in rote with a voice so young it almost seems cute. Then she smiles, swings her pink-stockinged feet in the air, and sticks a finger in her mouth. "*Muy bien,*" says the elder.

The next day I return to El Peregrino to talk with Joed Aguirre. His father, Tony, joins us on the patio. Joed is both a guide and a dive master who attended high school in the United States. At age 24 he has a classical face, dark wavy hair, clear brown eyes, and olive skin. Like his father, he believes that species adapt but that evolution is just a theory. "What scientists have shown is that animals change in a very short period of time. They haven't shown that the finch turns into a platypus or anything like that. They've shown that the finch changes, but he still stays within a certain genetic limit."

Joed believes that everything on Earth comes from an intelligent design created by a Master Worker. "In essence, Jehovah did design and place animals and living beings [here] as fully functional life-forms" within the last 6,000 years. He doesn't believe, for example, that birds were once reptiles. "I don't buy that science, no, because it's lacking in detail, it's lacking in the how of it, the steps by which you can trace it back without a doubt."

I mention all the evidence out there in the fossil record, in the radiocarbon dating of bones and antiquities, in the layers of geologic history in places like the Grand Canyon. It's just a theory, he repeats. There are too many gaps in the record. But not to most scientists, I remind him. I quote one, Gunter Blobel, a cell biologist at Rockefeller University in New York and a Nobel laureate.

"When it comes right down to it, you are not 20 or 30 or 40 years old. You are 3.5 billion years old. Some people may say how terrible it is, this idea that we come from monkeys. Well, it's worse than that—or better, depending on your perspective. We come from cells from 3.5 billion years ago. There is this tremendous thread of life that goes back to when the first cells arose, and that will continue on after any of us die as individuals. It's continuous life, and continuous cell division, and we are all an extension of that continuity. Reincarnation and similar themes are poetic representations of biological reality."

Ironically, Joed and other Witnesses deny that they're creationists. They point a finger instead at religions such as the Seventh-day Adventists, who literally believe the world was created in seven days. But the definition of creationism is broader than that. When I ask Joed if his faith ever collides with his work as a naturalist guide for the Galápagos National Park, his response contradicts what his father said earlier. No, he replies, he does not encounter problems in his work because he avoids talking about evolution. If passengers ask, he quotes authorities like geologists and biologists.

One of Joed's oldest friends suddenly appears, a "brother" named Daniel Fitter, who also works as a naturalist guide and photographer. Fitter joins the conversation. "Most people don't understand evolution," he says. If visitors ask him about it when he's out guiding he quotes others, even Darwin. "It's really not an issue," he insists. But it does present a paradox and a challenge.

Guides are often on tour two weeks at a time, and Joed says his main conflict is being away from the congregation. "As a Jehovah's Witness there's a responsibility, a lead that men take in teaching in the congregation. If I'm not here because I'm out traveling, that bothers me. But I do *not* find a conflict within the guiding job itself [while] believing in intelligent design and a Creator, and explaining biological settings here in the Galápagos Islands."

———

A few days later I meet with Washington "Wacho" Tápia, head of research at the Galápagos National Park headquarters, to ask about park policies on creationist guides. Tápia doesn't have much time this morning. People are running through the corridors as though the building's on fire. In a sense, it is. After almost 40 years of celibacy, Lonesome George, the last of his breed, has finally mated with both female land tortoises in his pen. Just yesterday his caretakers discovered a nest with eggs inside.

Everyone is so ecstatic the air crackles. I wait outside Tápia's office next to an avian veterinarian for the Charles Darwin Research Station, Gustavo Jimenez, whom I've talked to about bird monitoring projects. He and others have recently learned that a mosquito-borne virus called *Plasmodium* has infected the penguins in the western part of the archipelago. No one knows yet how serious it is.

Soon Tápia, a tall strapping man, opens the door and invites me into his office. Considering the hoopla over George, I'm surprised he's seeing me at all. When I ask about the guides, he seems perturbed. "They should not be discussing their religious beliefs. That is not what we contract them to do." Tápia admits that since the Special Law for Galápagos was passed in 1998 the quality of guides has plummeted. "First, there aren't that many prerequisites for becoming a guide," he says, like hard science. He goes on to tell me that one of the most "insipid" things is hiring a guide who does not love or understand nature, and that Galápagos schoolteachers must become professionals on an international level and understand why living on the islands is such a privilege. "This [failure of the educational system] is a reflection of Ecuadorian society," Tápia admits. The Galápagos can survive only if strong laws are passed and enforced to restrict development, if local attitudes shift so that people come to understand that this is a biological system unparalled in the world, and if the Galápagos school system is radically transformed.

"Education is the key."

On the way home from the park I stop at the Hotel Galápagos and ask Jack Nelson for his take on creationist guides. "It's not new. It's been going on here for a long time," he replies. "I wouldn't try to give you a number, but there are many small, assorted sects or churches that are fundamentalist, creationist, and in and of themselves *they* aren't the threat. The threat is the overwhelming ignorance of a population that can live in this place with all of the resources for knowledge and science and understanding and still believe in creationism. There is no way [that can happen] except by willful ignorance of facts or an inability to analyze and sort out the facts. It's a reflection of the very weak educational system in Ecuador."

I mention that during a recent visit to Quito I read in one of Ecuador's leading newspapers that Quito's schoolteachers had taken a routine exam, and 99 percent had flunked the section on reasoning. This comes as no surprise to Nelson. "They're not taught to ask any questions, or go out and find information, or to argue. Critical thinking is simply not part of the educational system in Ecuador."

Case in point: Nelson's 13-year-old daughter, Audrey, is just now studying the multiplication tables. Once she brought home a chart of the solar system that included Pluto as a planet. Even more alarming was that Io was also a planet. Io is one of the four moons of Jupiter discovered by Galileo Galilei in the early 1600s.

As for the park's practice of hiring creationists, Nelson says it's a product of the politicization of the park and its guide selection process. "If that guide meets the legal criteria—he's a Galápagos resident, he's a high school graduate, and he qualifies for the guide courses by memorizing the right question and the right answer—that's all that matters. I think it reveals the weaknesses of the system here and shows that a lot of the criteria are skewed.

"The Ministry of Tourism and the Chamber of Commerce have stacks of complaints from tourists about guides and boats and crews," Nelson continues. "So what? There is no follow-up. There is no authority. There is no payback even if the guide is accused of ignorance or arrogance or sexual harassment. Nothing happens. There are no systems for accountability. There needs to be a better feedback system of tracking on these guides. I don't know specifically if any fundamentalist or creationist guide has had complaints registered on his or her performance or adequacy. But I do know that if there *had* it wouldn't matter."

Few Galapagueños have felt the creationist whiplash more painfully than Carlos Valle, the Princeton graduate and evolutionary biologist I interviewed in Quito about the flightless cormorant. Valle says the newest threat to the Galápagos is religion, and he insists that the park should prevent fundamentalist guides from spewing dogmas and other inaccurate beliefs. "They cannot misinform or misuse their positions among tourists." But it does happen, as I learned on Bartolomé Island. Valle tells me that in the last five years, religious groups have found the Galápagos fertile ground in which to sow their seeds. Puerto Ayora alone is home to nearly 200 Jehovah's Witnesses, and schoolchildren learn creation science at the Seventh-day Adventist Loma Linda Academy, not far from the Charles Darwin Research Station.

Valle was born on Santa Cruz and grew up there before roads existed, back when the supply ship arrived once every four or six months and people bartered sugar for rice or beef. His mother died when he was eight and his older sister became the surrogate mom. "We are a close family, my brothers and sisters. We love each other." But Valle was unprepared for what awaited him the last time he went home to the Galápagos to spend a holiday at his sister's house. Half his siblings didn't show up. "I said, 'Why? What's wrong?'"

What was wrong was that half the members of his tight-knit family had converted from Catholicism to fundamentalism, including the Jehovah's Witness faith, whose followers aren't permitted to celebrate. "It's a sin," he says, shaking his head. "I asked them, 'What happened to the attitude that says I care about you, I love you? Jehovah is changing you.' I told my sister, 'Never be swayed by this type of religion because there will be no point, dear sister, for me ever to come visit my family in the Galápagos again.' " Valle squints, holding back tears.

"How is it possible for people living in the Galápagos, the mecca of evolution, to question evolution? In order to preserve a place like the Galápagos you need to understand evolution. If you reject evolution it's not possible to become a conservation advocate. The Galápagos needs every single citizen living on the islands to become conservation advocates. If not it means we'll lose the battle—and we'll lose the islands. It's absolutely contradictory in a place like Galápagos to have creationists and fundamentalists. It's not something you would expect."

CHAPTER 16

Puerto Ayora Redux

"Politics is the art of looking for trouble, finding it everywhere, diagnosing it incorrectly, and applying the wrong remedies."

—Groucho Marx

In June 2008 dozens of angry residents march through Puerto Ayora banging pots and pans in a protest called a *cacerolazo*. In Spanish *cacerola* means "stew pot." In some Spanish-speaking countries it's a popular demonstration by those who accuse the government of stealing their daily bread. Fed up with the injustice of being denied boat permits and watching those documents go instead to select fishermen and foreign tour companies is more than the protestors can bear. Among those marching are native Galapagueños and longtime pioneers like Jack Nelson and Steve Devine, who started tourism back in the 1960s by running horseback trips out of the Hotel Galápagos. Both men have dual U.S.–Ecuadorian citizenship.

As the throng of about 50 protesters marches along, chanting slogans, shopkeepers stand in their doorways and applaud. They, too, feel left out of what the Special Law for Galápagos, passed ten years ago, had promised them: more autonomy in their own province. Women carry babies or push them in strollers while balancing signs. A man with biceps the size of peaches bangs a cookie sheet with a rolling pin with such force that it crumples like aluminum foil. Two others carry a large banner that declares: "We want patents for all Galapagueños." Nelson stands out in a bright-red polo shirt. He's whacking a stainless-steel lasagna pan with a wooden spoon and carrying a sign that reads: "Tour de Bahia Operator since 1967." The goal of the *cacerolazo* is to make enough "noise" to embarrass authorities.

How did things get so out of hand? I wonder as I follow them down the street. After all, this is the Galápagos, that peaceful paradise—as the videos show—where equilibrium reigns. The fact is that the permit system to run a boat in the Galápagos is so Byzantine that anyone who hasn't lived or worked here would be completely confused by it. But that's the norm, Nelson tells me after the protest. "The central element of business in Galápagos is having a permit to run a boat. That's the gold mountain called a *patente* (patent), or a *cupo* (quota), a space." Those were selling for $30,000 to $50,000 a pop in the ministry in Quito back in the late 1970s to early 1980s, and very few people in the Galápagos could get one. "They didn't have the capital to go bribe for one or the right connections through the park to get one. Even though we started tourism here and ran a boat of our own, I went to the park with all of the documents to solicit a boat permit because they were handing them out at that time. The park *intendente* just laughed in my face and said, 'That was closed yesterday.' At that time I think 14 or 17 permits had been handed out. That got us organized; about 20 of us met for hours almost every night for two years." Finally in 1995, Nelson's bay tour and scuba diving group got congressional approval, but the regulations kept changing, and Scuba Iguana was forced to renew its permit on a regular basis. Under the current system the company is neither legal nor illegal. Today, there are about 80 permits throughout the Galápagos. "Some boats have more than one permit because they've managed to get them by mixing and matching and judicial legerdemain to combine five permits to run [big] live-aboard ships that can carry 80 or 90 passengers."

"How much does a permit cost now?"

"There are no new permits. Can't get one unless somebody wants to sell one on the open market, and it's not really open because it takes a lot of under-the-table stuff to make it happen. The Galápagos National Park policy—or theory—is that you can't transfer permits, and so it involves a lot of bogus contracts and things like that. But if you *can* get one, they're probably around $300,000 to $500,000—just for the permit. ... For those of us who live here, it's like being in prison because we can't do business. The real reason we're still operating is that there is no real authority in Ecuador because there is no rule of law. Tourism is *not* our resource. It belongs to the companies in Guayaquil and Quito. Somewhere around 95 to 97 percent of the tourist dollars never touch Galápagos."

Pioneers like Steve Devine are just as frustrated. Devine has lived in the Galápagos since he was four years old. His parents sailed out on a little boat from Seattle and his father found work as a civilian at the U.S. military base on Baltra Island. When the base closed, his family bought a small piece of land on Santa Cruz Island and began homesteading, hunting, fishing, and subsistence farming. When he was 14, Devine began guiding tourists on horseback from Nelson's hotel. Later in his small tour office, I ask Devine why the Galápagos have become such a political hotbed.

"You cannot have a park in complete isolation against the wishes of the local population. I'm sure there are a lot of park people and maybe a lot of conservationists who would like to see all of us colonists simply shipped back to the mainland and all of our lands confiscated." Today, all the park says is "no, no, no" for everything. "I can't even go out and walk on a moonlit beach with my wife. I can no longer camp at the beach on Tortuga Bay, and I've been going there all my life." The park closes the entrance to the beach at sundown. This started about eight years ago when they erected a fence with a locked gate after some kids were caught drinking and getting high.

"What do we have here in Puerto Ayora? The beach up at the Darwin Station, which is off-limits and is mostly rocks anyway. People need a place to go and have fun, to relax. Otherwise, why are they going to support the national park?"

Carlos Valle also feels the sting. "I can't even get a fishing license to bring a fish to my table, and I was born here." What's worse, he says, is that excluding the old residents and issuing boat permits to the same fishermen who burned down park buildings and butchered tortoises only "rewards the criminals."

It's that same irrationality that has residents so incensed today. "To give you an idea of how corrupt it is," Devine tells me, "I've been in tourism since before it [officially] started, and I have never been able to get a permit to operate a tour boat. I didn't have the connections and I didn't pay off the bribes. I didn't know how, or whom to bribe. It's a very sadistic method of dealing with the local population, the people who have acquired the right to be here."

So why stay? Once Devine tried to leave the islands, but like others he returned a few years later. "All my friends are here from my past. My parents are buried here. My kids were raised here. I was practically born here." He and his family still own some land in the highlands, where they run a few cattle. But, he laments, "our property taxes have gone sky high. You've got land that the municipality has valued at millions of dollars but is worth only thousands of dollars, and you're paying taxes on the millions of dollars. It's crazy. It's insanity. The problem is the volume of change here has been mind-boggling and chaotic. It's exponential."

The style of development is changing as well. The Royal Palm Hotel is a world-class resort up in the highlands, the only five-star resort in the islands where nightly rates range from $515 for a room to $1,201 for a villa. The owners even hired a young chef whose gourmet skills are among the best in the world. A few years ago real estate investment opened up in the Galápagos. Santa Cruz Gardens is a subdivision with 60 lots that allows its members to own a summer villa and to rent it out on a temporary basis. The subdivision has its own website, which boasts the names of celebrities who've visited the island: Robert de Niro, Olivia Newton John, Catherine Zeta-Jones and Michael Douglas, and Richard Dreyfuss. The contact for Santa Cruz Gardens is a company based in Yuba City, California.

Many residents believe that, in the end, populated islands like Santa Cruz may become sacrifice zones while the more remote islands may have a chance to survive. They have a point. In April 2008 about 50 people, backed by a dubious lawyer, attempted a Marxist land grab on Santa Cruz Island. Two groups of people calling themselves day laborers, farmers, and ranchers demanded that abandoned agricultural land be expropriated and divided among them, a total of about 560 acres. They accused impoverished landowners of not being real farmers and cited the Law of Agricultural Development, which states that expropriation is legal when lands suitable for agriculture have lain idle for more than two years.

Nelson calls the philosophy that supports land redistribution in the Galápagos "window dressing." Those who are trying to claim the land of others are almost never the landless poor, he adds, and certainly not on these islands. *El Colono,* the islands' semimonthly newspaper, devoted nearly a full page to the story. It quoted Galápagos Governor Eliécer Cruz demanding the detention of Jorge Cádenas, the lawyer representing these groups for "creating chaos" and interfering with "public order" among innocent people. He said the same people demanding the property had in some cases acquired the lands, then sold them. But Cruz did not stop there. He called their actions extortion against the Galápagos government, the respect for private property rights, and the established laws that prevent "invasions." The newspaper printed the full names of those demanding the land grab. A few residents have warned that if the battle for their lands continues, "there will be more than just pots and pans going off."

To be honest, the Galápagos could benefit from more agriculture. Santa Cruz has a food shortage, but it's not the result of laziness on the part of citizens. Clearing the available land would be a nightmare, especially where invasive species have left nutrient-poor soil. Equipment and labor are expensive. And then there's that nasty permit system. Most food is shipped to the islands on supply boats from the mainland, which is also problematic. The boats aren't fumigated and often carry invasive species, such as the Mediterranean fruit fly. During the first few months of 2008 a barge, the *Victoria,* nearly sank on the way to Puerto Ayora when a pipe broke and water began entering the hull. To lighten the load the crew threw the cargo overboard. Medical doctors who had shipped all their equipment to the town lost everything. Ecuadorian laws don't require ships to be totally insured. Because the demand for building materials like cinder blocks and heavy bags of cement is so great in the Galápagos, freighters are often overloaded. When they fuel up, the ships get so heavy they must leave cargo at the dock in Guayaquil. On the day the *Victoria* foundered, everything sank to the bottom of the ocean, including the town's food supply. A couple of weeks later a second supply ship developed problems and had to return to the mainland. Then a flotilla of million-dollar sailboats docked in Academy Bay and bought out the town. I'd heard a tale about Puerto Ayora's mayor when a similar food shortage hit a few years ago. The mayor stood on the dock, arms folded, guarding the village's beer supply. It was like an old West standoff as the mayor told tour operators, "You can buy our food but not our beer."

———————————

One day as I push a shopping cart through the main supermarket, which is next to the dock, I stop to examine the empty shelves. For several weeks now there's been no milk, juice, eggs, vegetables, chicken, or even toilet paper. I look up and see two women approaching. One throws her arms up in despair. "What is this, Russia?" she exclaims.

"No," I say. "Welcome to paradise."

I tell them about the boat problems and the food shortage. They tell me that the catamaran they booked a trip on months ago has developed engine trouble and they've been forced to cancel their eight-day tour of the islands. They've come all the way from Germany.

Boat mishaps in the islands are no joke. In 2001 an oil tanker, the *Jessica,* mistook a buoy for a lighthouse and ran aground, spilling more than 240,000 gallons of diesel and heavy bunker fuel into the pristine waters around San Cristóbal Island. The *Jessica* was on its way to deliver most of the fuel to the *Galápagos Explorer,* a luxury cruise ship—one of the biggest in the islands. Ecuador's President declared a national emergency while cleanup crews worked around the clock to clean up the spill and save endangered wildlife. Graham Watkins, executive director of the Charles Darwin Foundation, linked the accident directly to tourism. "There is an increasing probability of these kinds of accidents as a consequence of rapidly growing tourism and population on the islands," he declared. "This affects everyone, including those who depend on fishing."

Twice during my visits to the islands, tour ships have sunk while I was a passenger on a different boat—watching dolphins ride our wake, eating dinner, or rocking to sleep in my cabin. In September 2005 while I was traveling on the *Letty* with Ecoventura, the *Darwin Explorer* sank in Academy Bay, spilling diesel into the intertidal zone. In 2006 the boat I was traveling on with seven university students, the *Atlantida,* developed engine trouble and sent dangerous-looking black smoke into the sky. When one of my grad students tried to film the captain pouring water over the engine, he snapped, *"¡No fotos! ¡No fotos!"* The boat had only one rescue dinghy for 20 people. There were no life jackets, only minimal "floatation devices." Forget the radio. The communications system was broken. If the engine had caught fire all on board would have died. The boat floated in circles out in the Pacific late into the night while we huddled together to stay warm. As we drifted on the tide, our creationist guide from Bartolomé Island finally got enough charge on his cell phone to call for help. Two rescue boats arrived soon afterward and carried us to shore, shaken but safe.

It didn't stop there. In 2008 the boat I was touring on with 11 university students, the *Intrepido,* developed engine trouble and had to anchor in a place not authorized by the park while the crew radioed engineers. At the same time, maybe even on the same date, a boat just like ours, called the *Spondylus,* sank off Floreana Island. Passengers, some of them American tourists, clung to rocks out in the freezing water until rescue boats arrived. No one died, but one man suffered a bone fracture. In his role as U.S. consular agent, Jack Nelson was called in to investigate and to file reports on behalf of the American citizens on board.

In January 2009 the deluxe-class yacht *Parranda* caught fire and sank off Bartolomé Island. All 15 passengers, including two Americans, were rescued with only the clothes on their backs. The *Parranda* was owned by Quasar Nautica, which is considered one of the most reputable tour boat operators in the Galápagos. Only a few days later the company's website began offering tours on its other boats before an official investigation or environmental assessment could be completed.

To say that regulations in the Galápagos are unenforced or nonexisent is an understatement. In March 2007 the director of the Galápagos National Park, Raquel Molina, traveled to the island of Baltra to investigate an illegal tourism operation on Millionaria Playa that was being run by the Ecuadorian Navy. As Molina and her park rangers attempted

to seize kayaks and other illegal sports and fishing gear, naval person-
nel attacked and beat the park director and her rangers so badly that
some required hospitalization. Molina's neck was nearly broken; doctors ordered her to stay bedridden for two whole months.

Imagine this happening at Yellowstone or Yosemite!

After that event, the minister of the environment was called in on
behalf of the park. The minister of defense defended the Navy, which
retaliated by claiming the park director had not filed for clearance to
seize the Navy equipment on Baltra. Ironically, Galápagos National
Park rangers must get permission from the Navy before they go out on
patrol. This creates a catch-22, in that the Naval Port Authority always
knows the park rangers' whereabouts. This has made it much more difficult for the rangers to patrol the Galápagos Marine Reserve.

A few months after Molina recovered from her injuries, she was
fired. She was the 13th park director since 2002. Her misstep: refusing
to allow Roque Sevilla, executive president of Metropolitan Touring,
to combine numerous permits for his tour ship *Pinta*. As one of the
wealthiest men in the Galápagos, Sevilla has power, backed by Metropolitan Touring, the largest and one of the oldest tour companies in
Ecuador with more than 54 years of experience. Newspapers in Quito
reported that Molina was fired because she was "insubordinate" to the
minister of the environment. Molina countered by claiming that Metropolitan Touring was breaking the law. I tried to reach Molina but she
did not respond. So I paid a visit to Metropolitan's operations director
for the Galápagos, David Balfour, in his office on the other side of
Academy Bay.

"The bone of contention was the association of *cupos* (quotas),"
Balfour replies in a British accent. "The laws stated clearly that one
is allowed to associate two cupos to operate a vessel—two or more
cupos—it didn't state how many. This has been done many times in
the past." Molina insisted that only one permit could be used per

person per boat, but Metropolitan fought back. "We were trying to operate the vessel [the *Pinta*] with two cupos of 16 for 32 passengers," Balfour explains. "So we presented all the papers and she denied it for that reason." Metropolitan employed a lawyer who stuck to the old "legal" arguments. Molina again fought back, and soon the minister of the environment put the case in the hands of her own government lawyers. "They sat down and discussed it and said we were entitled to do it."

A day or so later the *Pinta* was legal to roll—and Molina was out of a job.

"I think what we're seeing here is the real mafia at work," Diego Quiroga later tells me at the Galápagos Academic Institute for the Arts and Sciences (GAIAS). Quiroga, a cultural anthropology professor and co-director of GAIAS, knows President Rafael Correa well. He was Correa's professor at the Universidad San Francisco de Quito (USFQ) before Correa ran for president. When Correa, a popular economics professor, demanded to teach full time while campaigning for the highest political office in the land, USFQ offered him only a part-time appointment. Correa retaliated. He sued the university, and won a settlement.

Meanwhile, the flap over illegal tourism and boat permits in the Galápagos remains as much of a "hot potato" as ever. In November 2008 all three mayors in the Galápagos and the leaders of three smaller towns wrote a scathing letter to President Correa, demanding a reform in the Special Law for Galápagos. The local leaders suggested that the president had been influenced by "dark interests" and had restricted the basic rights of permanent residents of the Galápagos. They demanded to meet with him in public to explain why he continues to treat Galapagueños like "an insular society." Whatever happened, they asked, to the guarantees and equity and inclusion, especially in the model of touristic development, "that during many years has been relegated to and has privileged the great economic interests of foreigners?" The mayors also asked that the president demonstrate the same patriotism he has shown for the similar causes he has promoted on the mainland. "To defend at least the less favored would signify that the confidence in you will not diminish in the heart of this pueblo that has made positive history and has remained sovereign for more than a century without even basic services for our children, young people, and families in general."*

---

*As of this writing, the president had not yet met with the mayors or other authorities in the Galápagos to begin talks.

CHAPTER 17

Into the Future

"Look deep, deep into nature, and then you will understand everything better."

—Albert Einstein

Our small bus climbs past quinine trees, Cuban cedar, and tall elephant grass—all invasive species thriving in the highlands just outside the capital of the Galápagos, Puerto Baquerizo Moreno. From time to time, a rancher appears on a skinny horse, and tethered cows graze on whatever's around. Soon the bus enters the garúa zone, where the mist is so thick you can catch it in your hand. I'm headed to the only freshwater lake in the Galápagos, El Junco, with 11 of my students, 3 park rangers, and Jorge Torres, a naturalist guide and former politician whose grandmother worked as a cook in the town of El Progreso just down the road. This is where the infamous Manuel Cobos enslaved, whipped, and murdered his workers until one of them retaliated by hacking him to death. The bus pulls over at the base of a hill leading up to the lake where we disembark.

Without a word, the wardens grab shovels, pickaxes, and several machetes and hand them around. Then off to work we go, like the Seven Dwarfs, our tools hoisted over our shoulders. The trail up to the lake is as slippery as butter, and several students glissade on their rears. I know the feeling. This is my third year helping the park eradicate thorny *mora,* or blackberry, and *guayabana,* both of which have choked out the native plants that once thrived above the lake.

The park workers are dressed in drab green jumpsuits and tall rubber boots. They have the thick hair and sharp features of those who have moved here from the Andes in search of a better life. At the top of the hill, the men give each of us rubber gloves. I show my students how to use a machete, then start hacking at a dense patch of mora. The men shake their heads in disbelief. Crazy gringos, they're thinking. Why would anyone want to do this nasty kind of work? I require it, I tell them. It's an important element of immersion journalism because students learn better through direct experience. Besides, invasive species are one of the greatest threats to this already endangered World Heritage site.

"Ah," Jorge Pallo says, nodding.

The other men get to work at once as the garúa turns to rain. Pallo has worked as a contractor for the Galápagos National Park since 2000, almost five of those years right here at El Junco. He has eradicated black rats and has monitored giant tortoises, Galápagos petrels, and sea lions. But nothing compares to this nightmare. Mora strangles what was once a lovely setting where visitors could overlook the lake and watch frigate birds swoop down over the water. What's happened here is a perfect example of how an ecosystem can go down from a few tiny seeds. "The hardest work of all is here at El Junco," says Gallo. For one thing, the weather itself is a double-edged sword. In 2007 Pallo and his crew planted 1,500 endemic miconia bushes, but it was a dry year and a thousand of the plants withered and died. On the other hand, too much rain pumps life into competing blackberries. The park once used the herbicide Roundup, but the wind blew it down toward the lake so the workers returned to machetes.

Then a new alien appeared in the lake when someone illegally stocked the pristine waters with tilapia, a fish from Africa. The park brought in a plant from the Amazon called barbasco. Resin from the root is used as a commercial insecticide and piscicide, or fish poison.

One of the major components in the plant is rotenone, a natural but highly toxic substance. The tilapia are believed to be gone, but the blackberries? Maybe never. After a few hours of work 16 of us have cleared and removed so many blackberry plants that the heap rises above our heads.

So how did such an invasive species get here in the first place? "If a bird like a finch eats the fruit and poops when it flies away, the seeds sprout in the soil," Pallo explains. "The people who brought this plant from the mainland had no idea how aggressive or successful it is because in Galápagos the land is so fertile."

It will take another eight years to eradicate all the mora around El Junco, maybe five, he says, if the local farmers become conscious and start controlling these plants on their own land. I admire his optimism, but it seems inconceivable that this fortress will ever return to its original state. While stooping to chop a rather big plant, I remember something from a fairy tale and remark: "This is like trying to empty a river with a slotted spoon." Unless the roots are yanked completely out the plants return like Lazarus.

Noon comes and goes and we hike down to the awaiting bus, much more confident now as we carry our machetes. We've braved the cold, the wind, the rain, and those infernal thorns that pierced through our gloves. But in doing so we've given something back to these endangered islands.

A few miles down the road three tall white turbines rise against a mountain of green, their blades whirling through mist. This is the San Cristóbal Wind Project, one of the most promising examples of renewable energy in the Galápagos. Wind power entered the arena of solutions to the many woes here in 2001 after the oil tanker *Jessica* crashed right offshore. The wind project is designed to eliminate diesel fuel completely by the year 2015. During the windy months from October through December, these three turbines can provide between 60 to 80 percent of the town's electricity, according to José Moscoso, director of operations for the $10 million project. Wind power will prevent oil spills because fewer tankers will be needed for the burgeoning tourism industry. Air pollution and greenhouse gases will also be greatly reduced. The hybrid wind-diesel project was made possible by the Ecuadorian government, the United Nations Development Program, and the nine largest electricity companies in the world. But like most things in the Galápagos it faced some serious obstacles, including the need to avoid the migratory path of the endangered petrel.

At first engineers wanted to build the wind farm on a mountain called San Joaquín, the highest peak on San Cristóbal. But biologists said no. The project would further harm an endangered species. Amid much turmoil, the turbines were installed on a different peak: Cerro Santo Tomás. To see and hear these windmills gives one a sense of hope for the future. The turbines run on a network of computers controlled by satellite all the way from Spain, according to José Jara, a young operations and maintenance technician for the project. The good news is that San Cristóbal's wind farm is so promising that a second site is planned for Baltra Island. This unit will supply energy to Santa Cruz, the island with the largest population and the greatest energy demand. Environmental assessments show that Santa Cruz is much windier than Baltra, but too susceptible as home to countless endemic and endangered species. The ultimate goal is to provide renewable energy to all the inhabited islands in the Galápagos. Other solutions like wave power from the ocean aren't feasible in the protected marine reserve. And in a bioregion where volcanoes still give birth, geothermal energy has yet to be tapped because of restrictions in the park.

If alternative energy is the way of the future, so is recycling. On a sunny June morning, workers on the conveyor belt at the San Cristóbal recycling plant are swearing under their breath because the largest tour boat in the Galápagos—the *Explorer*—has just dropped off a truckload of garbage. The ship's crew has failed to sort it, and now the workers must rip open each bag and separate the glass from the plastic from the paper and so on. Tour boats are by far the largest polluters in the Galápagos.

Jaime Ortiz, director of the project and an environmental director for Puerto Baquerizo Moreno, is happy I'm here to witness this infraction. Recycling is now mandatory here in the capital of the Galápagos and everyone, including the big tour companies, must comply. Every week the center receives about 440 pounds of glass alone, mostly Gatorade bottles. Ortiz plans to send a warning letter to the company. If the *Explorer* still doesn't comply it can lose its environmental certification and be fined up to $200, which many consider a slap on the wrist.

But Ortiz and others hope that recycling will become a model of sustainability for the locals. Each family has been given three bins in which to separate glass, plastic, and organics. If they don't comply they receive a visit, often from high school students whose task it is to educate them. The town dump, they're told, is no longer a solution. The trash there must be incinerated because digging into the hard lava is too difficult. Much of the debris is toxic, including medical waste. The smoke from plastic alone can release persistent organic pollutants, or POPs, including cancer-causing dioxins. The leftover ash is also toxic and must be transferred elsewhere.

Despite the new recycling plant, a stroll through this town reveals trash everywhere. If recycling is mandatory, I ask, why are the streets so littered?

"It will take time for people to get used to the idea of recycling," Ortiz replies. "Until then, they'll keep littering."

That afternoon at GAIAS I ask Diego Quiroga if he thinks recycling will eventually catch on here among the locals. "Maybe in about ten years. To them there is no difference between a plastic bottle and an orange peel."

Santa Cruz Island, which is about three hours away by speedboat, also recycles waste, but it's not mandatory there. The center was made possible by grants from numerous companies and nonprofit organizations such as Wild Aid and Fundación Galápagos. Metropolitan Touring runs the project. Waste is shipped to the coastal city of Guayaquil and sold to recycling companies. Plastic, I'm told, is sent to China to make sweaters. But it costs Metropolitan about $2,000 just to send one shipload of recyclables to the mainland 600 miles away, despite Ecuador's fuel subsidy.

―――――――――

And what about those unleashed dogs in Puerto Ayora that claw into bags of trash placed on the ground because people can't afford plastic recycling bins? One day I take a taxi to CIMEI, the animal control center north of town, to see why. The acronym translates to Inter-institutional Committee for the Management of Introduced Species. It's part of the municipal government and includes the Charles Darwin Foundation, the Galápagos National Park, public health officials, the Environmental Police, and nongovernmental organizations like Wild Aid and Sea Shepherd. But it's not just cats and dogs the group monitors in the urban areas of the archipelago, Freddy Salas tells me from behind his desk. "Our job is also to eliminate plagues like cockroaches, wasps, yellow jackets, rats, and mosquitoes, especially those carrying dengue." Household pets must have a picture ID and a chip inserted into their necks.

Salas believes each family in the Galápagos should be allowed a single pet but that they must follow the regulations: dogs must be leashed, tied up, or restrained behind a fence. Sarah Darling agrees. Darling, an artist who runs a gallery called Angelique in Puerto Ayora, can often be seen walking around town in a dress with her adopted dogs, Chocolate and Neptune. She believes that owning a cat or dog is good for humans. "It helps encourage a saner, healthier lifestyle and it alleviates stress. It has been proved that dogs taken to hospitals to visit the elderly or infirm can work miracles. I say yes to owning a dog or cat in the Galápagos as long as the owner treats the pet like a member of the family. People would not allow their children to chase iguanas. Likewise for their pets." Darling also believes that the rules here are too lax. "It should be imperative that all dogs and cats be registered and sterilized," she says, and that anyone caught smuggling or breeding dogs should spend six months in jail.

Ironically, in a place like the Galápagos it's not mandatory to sterilize one's pet. It's optional in Puerto Ayora, where 55 percent of the dogs have not been spayed or neutered. Salas says that in 2007 CIMEI euthanized 185 dogs and 163 cats. "If I find them in the street I capture them and sterilize them," he says, swiping his hand through the air as though catching a fly. Now there's a new money-making scheme on the island: puppy mills. People with money are smuggling in *perros de raza,* pedigrees, to breed as a type of business. Among them are poodles, daschunds, great danes, mastiffs, pit bulls, spaniels, and bull dogs. "It's illegal, but it's *not* illegal because it's not yet in the law," Salas laments. CIMEI is planning to build a small DNA lab to determine if a dog was born in the Galápagos or smuggled in. If the latter is true, the animal will likely be euthanized. Every year the international group Animal Balance comes to the Galápagos to help local authorities with sterilization programs for cats and dogs, as they do on other fragile islands.

Even so, Salas is visibly frustrated. Not long ago the big cargo ship *San Cristóbal* was so overloaded it had to leave food on the dock in Puerto Ayora. "I know this is how the Mediterranean fruit fly got here.

I saw worms and flies in crates of food left on the dock. It's so obvious. I'm not a scientist but I'm not stupid either." Salas insists that for a boat to enter the Galápagos it must be fumigated and pass a rigorous inspection. "Sadly, that is not the case. My hands are tied because I don't have the authority to do anything."

When I leave his office I pass a CIMEI veterinarian and two cages with kittens that hiss, recoil, and try to claw her through the wire. "These cats are not for adoption," she says, frowning. "They're really feral cats. I can't even get near them they're so wild."

———————————

Godfrey Merlen, who wears his hair in a long gray ponytail, has worked most of his life with the Galápagos National Park to help guide and strengthen the organization. As the director of the nonprofit group Wild Aid in Galápagos, he and his co-workers try to break down social barriers and influence the way people see things. That role is critical in the long-term survival of the natural resources. "If the resources go down, everything goes down," says Merlen. "If the resources go down, tourism will go down." There isn't a single person in the Galápagos who doesn't profit from the natural resources, he adds, including fishermen. "Since the park is responsible for those resources, it must fulfill its role." But widespread corruption has made it nearly impossible to protect the islands.

Ever since Merlen came here from England in 1970 he has worked with other sectors to establish mutual trust. Wild Aid has even brought in rock bands to reach young people, and it runs an ongoing campaign to educate fishermen. "It's common knowledge that there is a constant trade in sea cucumbers, in shark fins, in even sea lion penises. Apart from being just brutal and ruthless, it's really dangerous," he says. "You're cutting the genetic core, damaging the essence of a sea lion colony by taking out the big males. You're hacking away at what

Galápagos stands for, which is the adaptation and the evolution of animals under certain environmental restraints."

Wild Aid helped the park secure two sniffer dogs, Labrador retrievers trained in Napa Valley, California, to sniff out drugs, shark fins, and sea cucumbers. In 2006 one of these dogs, a black Lab named Aggie, helped the Environmental Police detect over 1,400 sea cucumbers hidden illegally in boxes and ready to be shipped off the islands. This major bust would not have been possible without Wild Aid's help. More recently, Merlen convinced the two Ecuadorian airlines that fly into Galápagos to fumigate incoming planes. Many tourists are coming from the Amazon, and the odds are excellent that unwanted hitchhikers might be on board.

I run into Merlen again on Earth Day 2008 on the dock at Pelican Bay, where environmental groups have set up booths. Here, against a backdrop of mangroves and diving boobies, Merlen puts an iguana puppet on his hand, squats in front of some children, and talks to them about nature. Members of Ambiente Independiente (AI) (Independent Environment) are also here to educate the locals. AI is a coalition of mostly women who volunteer time between their day jobs and child-rearing. The main goal is to educate the community about reusable objects and to call attention to the colossal waste from plastic bags. A poster declares that every single minute the world uses one million plastic bags. That's enough to encircle the planet 63 times. One of their most impressive campaigns, *Fundas Lógicas* (Logical Bags), inspires people to sew and adorn reusable nonplastic bags with personal slogans. Two of my favorites are: "*Sí, la solución está en mis manos*" ("Yes, the solution is in my hands"), and "*No soy una chica plástica*" ("I'm not a plastic girl"). The group also promotes bicycle riding, organic vegetable gardens, and native plants.

That's good news for FUNDAR–Galápagos (Foundation for Alternative Responsible Development in Galápagos), which is run by Carlos Zapata. The nonprofit group is here today on the dock with organic produce from its experimental farm: leafy green basil and two varieties of lettuce, perfect onions, carrots, broccoli, and long red chili peppers. FUNDAR has erected a food dehydrator made of 100 percent recyclable materials to demonstrate what's possible. Nearby a natural insect trap set up on a homemade tripod draws flying bugs to a flame, then drops them into a pot of water to drown. I'm impressed with what I see, and I plan a trip to FUNDAR's demonstration farm in the highlands of Santa Rosa, home to about 700 people.

At a modest building with a sheet-metal roof, Martin Espinosa, an engineer with salt-and-pepper hair, and Ivon Aldaz, a bespeckled biologist, greet me. The nonprofit group depends on volunteers, and this week 16 students have come with their teachers from a high school in Berkeley, California. FUNDAR works with the local community on sustainable development, alternative and renewable energy, environmental protection, water management, and agroecology. Its goals are to create a new paradigm that integrates conservation with environmental ethics. FUNDAR wants to turn the Galápagos into "a model of harmony between humanity and nature" and stop the antagonism between certain sectors. The group offers scholarships, even on the international level.

The property, about 208 acres, sits on a former cattle ranch where invasive species still hold fast, including elephant grass, quinine and cedar trees, rats, and nasty fire ants that sting my ankles as I follow Espinosa and Aldaz down a fern-lined trail to the farm. Rain pelts the ground like hailstones. At times I must step over the exposed roots of cedar trees, hosts to pale green lichens and shelf mushrooms. Then as I reach for a branch, I slip and fall in the mud. Finally we come to the organic garden, a little piece of heaven on this parched and rocky island. Rows of lettuce and tomatoes glisten in the rain. Radishes, cabbage, cauliflower, and broccoli thrive in the organic soil. Barriers of natural insecticides control pests: chili peppers, garlic, and marigolds.

FUNDAR uses drip irrigation from collected rainwater. Six cows and some chickens provide precious manure, and a greenhouse contains 40 species of endemic plants earmarked for farmers. Sloshing back up the trail, I realize that the coolest thing FUNDAR does is to bring people together, including about 50 Galápagos farmers who attend long-term workshops on organic farming. On Santa Cruz that's progress. As Peter Kramer, President of the CDF Board of Directors stated, "Galápagos is a local treasure—and it can be preserved only with and through local people."

Kramer is absolutely right. Local involvement is key in a place where social resentment toward outsiders remains strong. But things are changing, albeit slowly. On International Coastal Cleanup Day 150 volunteers and 35 divers collected almost 2.5 tons of trash from the beaches and seafloor around Santa Cruz Island. Imagine that much litter and the threats to the islands become all the more tangible.

On the scientific front huge gains in preservation have been made as well. Sea turtle populations are once again healthy. Sharks are now being tagged and monitored to learn where they migrate. Tortoises and land iguanas have been repatriated to their native islands. Goats, donkeys, pigs, and rats have been eradicated from several islands. Some species are now being analyzed through their DNA, and the CDF is awarding scholarships to Galápagos natives who want to pursue a career in scientific research.

Despite the strides, everyone agrees that environmental education is conspicuously absent in the schools. One day while my housekeeper

Francisca works her way around my house, I move to the bedroom and read the December 2007 issue of *National Geographic* magazine. The cover story is about dinosaurs. The issue includes a foldout that identifies perhaps a dozen species that are mentioned in the text. I've never had a housekeeper in my life, nor have I seen anything to equal the small lake she creates as she mops my floor. Francisca is a Salasaca Indian from the Ecuadorian highlands and today she has brought her five-year-old daughter, Micaela, who stands there at the foot of my bed staring at me. Her soft brown eyes never once lose contact with mine, and I pat my bed to invite her up. As I show her the dinosaurs in the magazine, she points to the map to identify each species. *"¡Este!"* she exclaims proudly, matching each image perfectly. "This one!" On another day while Francisca is sweeping, a baby gecko flees from beneath my couch. *"¡Un lagarto!"* she exclaims, "a lizard." *"Sí, un geco,"* I say, scooping it up in my hand to take it outside.

"Aren't you afraid?" she asks.

"No, geckos don't bite. And they're really cute."

"Go show Mica," she insists, waving me out the door where Micaela is swinging happily in my hammock.

"Mira," I say, opening my hand. "Look."

Micaela's eyes fill with wonder at the living treasure. Then the gecko jumps from my palm and lands on the volcanic cinder in my patio where it remains motionless.

"Dead or alive?" I ask, testing her.

Micaela stoops to touch it, without fear. The startled reptile scrambles away, much to her delight. To me, this is proof that children contain an innate wisdom, a natural sense of curiosity, and I'm saddened to think that Micaela's perfect sense of reasoning might soon be twisted by an antiquated school system where most teachers are far less wise.

Not long afterward I take a water taxi to a tiny bay called Playa de los Alemanes with Ivonne Torres, one of the longtime naturalist guides

I interviewed earlier. Her two daughters have come, and as we follow a trail to the cove one of them says in a tiny voice, "Mommy, we don't have any toys to play with."

"My darlings, you don't need toys at the beach," she says, putting her arm around the girl. "You can play with rocks, or seashells, or build sand castles."

Shania is six and Shaiel is four. They look like twins as they walk hand in hand down to the shore, the sun sparkling in their long dark hair. One of them finds a Styrofoam cup and the two begin filling it with water and the empty shells of brine shrimp. I sit in the shade of a black mangrove tree with Torres, watching. "They think they're alive," she says, laughing. They're being so careful not to harm the translucent entities floating in the cup.

But soon they grow weary of this game and invite their mom out into the water. Torres agrees, but only if they collect litter on a small stretch of the beach. She hands Shania a plastic bag and the girls get to work. When they return Torres kisses them and takes each one by the hand into the water where she points out a lava heron feeding its chick and a pelican perched stoically on a rock. I remain on shore, deeply moved at the immense possibilities before me. This level of participation is exactly what the Galápagos needs to survive.

The sun dips low and soon it's time to leave. Shania has draped her beach towel over a branch on the mangrove tree. To dry off she spins around in it until her torso is completely wrapped. "I'm a cocoon!" she yells. Then she spins the other way and emerges with outstretched arms, face skyward.

"Now I'm a butterfly!"

Diego Quiroga believes it's never too late to turn things around, even though education is an uphill battle in the Galápagos. But parents must begin teaching their children early—reading to them, exposing them to nature in the raw, and sharing in the fun. On any given day on San Cristóbal Island, Quiroga, a professor of cultural anthropology, heads into the clear blue ocean in a Speedo bathing suit and swims for miles. The fit and handsome 45-year-old is a former Olympian. At the age of 12 he was a member of the Ecuadorian national swim team, at 15 he was the South American champion, and he later participated in the Pan American games. Quiroga's father, a glass etcher from Spain, had wanted his son to become a marine biologist but could not ignore his talents. He sent him instead to Ft. Lauderdale, Florida, to train with a famous coach named Jack Nelson (not the same Jack Nelson of Scuba Iguana). Later, At U.C. Berkeley Quiroga trained with U.S. Olympic swimmers, and in 1980 he represented Ecuador at the Moscow Olympics.

But Diego Quiroga is more than a sterling athlete. He's a visionary—one of the founders of the Galápagos Academic Institute for the Arts and Sciences (GAIAS), a satellite campus of Ecuador's leading academy, the Universidad San Francisco de Quito (USFQ). GAIAS opened in 2002, partly at the request of the islanders, who were tired of distance learning and who wanted their own academic center on San Cristóbal Island. Quiroga, along with evolutionary biologist Carlos Valle and professor of marine fisheries Gunther Reck, created the satellite campus.

Reck, who is German born, was director of the Charles Darwin Research Station from 1984 to 1988 when discussions led to the creation of the marine reserve. He worked closely with traditional fishermen and was an original promoter of the Special Law for Galápagos. He was also an adviser to the minister of the environment. Reck wanted to assure that the fishermen would be the real champions of the resources they depended on. The next thing he knew the fishing cooperatives had doubled in size, thanks in part to Eduardo Véliz in the 1990s, and no one knew who was legitimate and who was not. "I'm now aware of the hundreds of loopholes in the Special Law," he says. "At the time, it was something very good, something fantastic with the consensus of the whole population, but obviously I was naïve." That's why the law is now under scrutiny for revision.

GAIAS offers semester-long programs, taught in English, to Ecuadorian and international students. Among the main goals are to educate islanders and train them in skills that are not damaging to the environment. One program is designed to help the park train new guides. To qualify, students must complete rigorous classes in ecology, marine biology, and geology. Other courses explore solutions to the social, political, economic, and environmental problems on the islands. And now GAIAS offers a master's degree in marine biology. The joint programs of USFQ–GAIAS support one of the most extensive exchange programs in the world.

But for locals GAIAS is more like a community college where students follow a two-year curriculum leading to an associate in arts degree in tourism, environmental management, or business administration. Recently, the World Wildlife Fund, which is closely aligned with the CDF and Galápagos National Park, gave GAIAS major scholarship funding. The newest push in education comes with an alliance between GAIAS and the University of North Carolina to establish a research institute for marine biology in Villamil. The institute will also promote sustainable development on Isabela Island. But professors like Reck warn that to avoid further erosion of the ecology, the population must be limited.

There are curious ways to make a point, and Diego Quiroga and Galo Yépez, the first Ecuadorian ever to swim the English Channel, decided their swimming ability could be useful. In 2005 they made headlines by swimming the seven and a half miles to Puerto Baquerizo Moreno from the famous dive site León Dormido (Sleeping Lion), also known as Kicker Rock. A year later they made an even more daring journey through shark-infested waters while a rescue boat followed. Why would anyone as bright as these two attempt such a feat? To raise awareness among islanders and mainland Ecuadorians about the degradation of the islands and their wildlife—especially the slaughter of sharks. Wild Aid and the World Wildlife Fund helped with the campaign. Quiroga had talked to marine biologists about swimming in open shark-infested waters, and they told him the sharks wouldn't pose a problem. Quiroga is a dive master who has swum peacefully underwater among sharks for many years, and he didn't stress. But others did, including his wife Tania and Gabriel Idrovo, the doctor who runs the decompression chamber in Puerto Ayora. In that campaign, called Swimming for Galápagos, Quiroga and Yépez made a 50-mile swim between the islands of Santa Fe and Santa Cruz. No one had ever accomplished this feat, nor had anyone dared to try.

Two days before the team was scheduled to set out they had to postpone the event because of a jellyfish invasion. "I'd never seen it like that before. It was literally a soup of jellyfish floating around. If we had gone swimming that day we would have been killed," Quiroga recalls. A few days later the waters cleared but not entirely. "Every five or ten minutes I was stung by one but it was manageable. You get used to those things."

Television crews from Quito and Guayaquil covered this astonishing event. It gave Quiroga a chance to talk about the problems in the Galápagos on national television and to highlight the dangers threatening the Galápagos for average Ecuadorians, most of whom cannot afford to travel here. "It's difficult for people who live [on the mainland] to understand the concerns and complexities of the Galápagos."

Quiroga, who now has a young son, is saddened by the lack of contact with nature among Galapagueños. He can't understand how people can live beside the ocean but not enjoy it. "They'd rather go and drink beer in a bar instead of having fun snorkeling. It's very difficult to get a 40-year-old man to start snorkeling. But maybe if you start teaching kids when they're small they'll enjoy the ocean and feel safe in it. We should teach people to have fun in nature. But I don't think that's here yet in this culture." Lately, Quiroga has seen a positive change among youth on the islands: Surfing. "More young kids are learning how to surf and are feeling more comfortable with the ocean, but it's a slow process."

That brings us to the ultimate question: Can the Galápagos survive? A few years ago Quiroga attended a workshop with two professors from Spain and three from Ecuador, one of whom was Gunther Reck. The scholars came up with three distinct scenarios for the islands' future. They invited people from the Charles Darwin Foundation, the national park, and the governor of the province to sit in. Here are their theories:

Scenario One: The Galápagos will be well preserved and the population maintained at 30,000 to 40,000. Money and investments will be controlled and the ecoregion managed as impeccably as possible.

Scenario Two: The Galápagos will become a "techno-zoo," a big park like Busch Gardens in Florida, where wildlife is managed and controlled like "a nice fantasy island." Not with big chain hotels or luxury resorts but with bungalow-type hotels where people can feel at one with nature. "Maybe you can have 200,000 or 300,000 visitors here [a year], but not two million," Quiroga adds.

Scenario Three: The Galápagos will welcome three million tourists a year, build a golf course on Isabela Island, put in big hotel chains, allow sportfishing, and promote para-sailing from yachts that tether and pull tourists on parachutes as they do in tourist towns like Puerto Vallarta in Mexico. "People will come here for the sun like the TV show when college kids go to Cancún for spring break and get half naked and drunk." That kind of tourism, he says, is "the nightmare of conservationists."

Some people at the workshop worried that all three scenarios are possible right now. Others said it's already too late for the first option— the Galápagos as a world model of conservation. "Unless we make some drastic changes we'll never get there. It's like when you're on a highway and you're close to the exit, but you're going too fast and you pass it and there's no return," says Quiroga. "When we pass that exit, the next one will be the techno-zoo."

Even so, Quiroga believes the Galápagos tourism model has not been a complete failure on the more remote islands. The towns, however, are another story. Santa Cruz Island, he says, has become a complete disaster, a sacrifice zone. "There are too many people. Let's be realistic, though. You will not be able to kick people out of Santa Cruz. What we should do is make sure nobody goes to live on Santiago or Española. The Galápagos will always be at risk. The new battle cry is to save what's left."

Epilogue

In the end it's the human species that holds the key to this "last paradise," the only oceanic archipelago in the world that retains 95 percent of its original biodiversity. If the 50 years since the Galápagos National Park was born are any indication of the future, then radical measures are needed at once to halt the onslaught of tourists, colonists, and invasive species. A major study published by the Charles Darwin Foundation and the World Wildlife Fund, "A Biodiversity Vision for the Galápagos Islands," supports this notion entirely. If political chaos continues at the current rate, the loss of species will become irrevocable.

I cannot imagine a Galápagos environment where the wildlife stops trusting us, where we can no longer snorkel with sea lions, turtles, penguins, and hammerhead sharks. I cannot imagine cliffs devoid of flightless cormorants, or frigate birds, or red-billed tropicbirds. I cannot imagine an ocean emptied of sharks or lobsters or whales by human ignorance and greed. Nor can I bear losing the privilege to watch with these human eyes the courtship rituals of blue-footed boobies stomping the ground with their quirky feet, or an albatross pair bowing to each other or clacking beaks like a drawn-out kiss. Something will truly be lost if the day ever comes when a human-induced oil spill wipes out the very species tourists come to see. That, say the experts, is a disaster just waiting to happen.

The Galápagos Islands need help, and lots of it. But preserving these sacred islands cannot work without a rigorous education system where resident children and their parents learn to coexist with nature instead of exploiting it. The same is true for the large tour companies, whose impact here sets everything in motion—from the fuel used in the airplanes that carry nearly 200,000 visitors a year, to the millions of camera batteries that must be recycled as hazardous waste. The human footprint is huge.

Without the long-term commitment of international groups much of this natural wonder would have vanished long ago. The islands have survived precisely because billions of dollars have poured in from non-profit groups like the Charles Darwin Foundation, whose role is to advise the Galápagos National Park on wise management through science and education; the Galápagos Conservancy; the World Wildlife Fund; Wild Aid; Sea Shepherd; Conservation International, and hundreds of others too numerous to mention.

Change is inevitable but never have I seen a decline as rapid and shocking in scope as during the time I spent living in the Galápagos. Often it was difficult to keep up with all that was happening, and at times, impossible to track. Here's a list of the changes I observed in less than a year:

- Raquel Molina, the 13th director of the Galápagos National Park since 2002, was fired.
- Graham Watkins, Executive Director of the Charles Darwin Foundation, resigned to pursue research elsewhere.
- Galápagos Governor Eliécer Cruz resigned after a year in office and was replaced by Jorge Torres, a former politician who one day worked side by side with my students and me to eradicate invasive blackberries.
- Juan Chavez, the Galápagos National Park director on Isabela Island who hid out in the mangroves after fishermen torched his house, resigned.

- Martha Véliz, who is married to Juan Chavez and who taught in Villamil's environmental school, also resigned.
- Eduardo Véliz, the Galápagos congressman who instigated the violent fishermen uprisings and changed the course of history in the Galápagos, returned unexpectedly to the islands in early 2009, some believe to run again for office.

The fact is that this game of musical chairs in a politically unstable province is a major reason the islands became endangered in the first place. The Charles Darwin Foundation stated in a recent report, "Galápagos at Risk: A Socioeconomic Analysis," that UNESCO's decision to include the Galápagos on its List of World Heritage sites in Danger "offers the best and last opportunity" to assure the future of conservation by creating a sustainable society.

Given our history as a species and the impacts of climate change, the Galápagos will forever remain at risk, but nature always wins in the end. Long after we humans are gone—to the moon, to Mars, to the dust we came from—blazing new islands will push up through the waves over the volcanic hot spot near Fernandina Island as the older islands to the east return to their home on the ocean floor. It's a cycle that has continued in the Galápagos now for 90 million years. Some species will continue to adapt and evolve in this brave new world and others will vanish. Those that survive may morph into creatures that look nothing like their ancestors. Life will continue—resilient as ever.

Poetry is one of my earliest loves: Each word must resonate like a finely tuned flute. Pattiann Rogers is a poet whose vision captures the essence of the natural world with insight and passion—like a primal instinct—from the soliloquies of whales to a tiny gold speck on a butterfly's wing.

The following poem, "A Statement of Certainty," resonates with grace and wonder at how life continues to evolve in strange and lovely ways we may never fully grasp. Therein lies the magic of certainty: Change.

A Statement of Certainty

Here we are, all of us now, some of us
in emerald feathers, in chestnut or purple,
    some with bodies of silver, red,
    or azure scales, some with faces
of golden fur, some with sea-floating
sails of translucent blue, some pulsing
    with fluorescence at dusk, some
    pulsing inside shell coverings shining
    like obsidian, or inside whorled
and spotted spindle shells, or inside
leaves and petals folded and sealed
    like tender shells.

Because many of us have many names—
    black-masked or black-footed or blue-
footed, spiny, barbed, whiskered or ringed,
    three-toed, nine-banded, four-horned,
whistling or piping, scavenger or prey—
we understand this attribute of god.
Because some of us, not yet found, possess

no names of any kind, we understand,
    as well, this attribute of god.

All of us are here, whether wingless
clawless, eyeless, or legless, voiceless,
    or motionless, whether banging
as pods of fur and breath in branches
    knitted over the earth or hanging
from stone ceilings in mazes of hallways
beneath the earth, whether blown across
oceans trailing tethers of silk, or taken
off course, caught in storms of thunder
currents or tides of snow, whether free
in cells of honey or free over tundra
plains or alive inside the hearts of living
trees, whether merely moments of inert
    binding in the tight blink of buried
eggs, or a grip of watching in the cold
wick of water-swept seeds, this—beyond
    faith, beyond doubt—we are here."

—*Pattiann Rogers,*
from Generations, *2004*

Acknowledgments

Galápagos at the Crossroads endured a long gestation period during my many visits to the islands as a writer and professor, especially during the eight months I lived on Santa Cruz Island between December 2007 and August 2008. My goal in writing this book was to immerse myself in the field like a cultural anthropologist, to capture pieces of ordinary life in an extraordinary region, and to tell a story about change. Through keen observation, I became intimately familiar with people in their own culture and environment. I tried to understand and synthesize widely differing views.

The majority of this book includes enterprise reporting in the field. I was especially interested in capturing the stories of Galápagos natives and early pioneers like naturalist Jacinto Gordillo, a former priest who at age 84 still raises endemic and endangered plants in the highlands of Isabela Island. All scenes unfold exactly as I observed them, and all historical events are reconstructed based on facts and the memories of those who lived through them. In most cases I taped my interviews and had them professionally transcribed. During the eight months I lived in Puerto Ayora I relied on the Charles Darwin Research Station's library, one of the best and most extensive resources in the Galápagos. Bookstores are almost nonexistent there, and one of my main challenges was the absence of reliable information. As a result, I had to retrieve a great deal of data over the Galápagos internet system, that is, when it was working.

So many people helped make this book possible that it would be difficult to name all of them. First I'd like to thank Paul Primak, director of Oregon University System International Programs who in 2005 invited me to create an on-going program called "Environmental Writing in the Galápagos," and Surendra Subramani, who recommended me in the first place. The University of Oregon Office of International Affairs has been enormously helpful over the years. Many thanks, too, to Dave and Nancy Petrone for the Petrone Faculty Fellowship "for research or teaching that advances communication in the public interest." Their gift provided the resources I needed to design this overseas program and begin research for a larger project. I'm also grateful to Tim Gleason, dean of the School of Journalism and Communication at the University of Oregon for allowing me sabbatical leave to write this book.

My colleagues at the Universidad San Francisco de Quito (USFQ) and the Galápagos Academic Institute for the Arts and Sciences (GAIAS), a satellite university in the Galápagos, taught me more than I dreamed possible: Diego Quiroga, Carlos Valle, and Günther Reck. Diego Quiroga is a co-founder and director of GAIAS, who has accompanied me into the field many times during my summer program. His background as a cultural anthropologist and his interest in the relationship between people and nature in this UNESCO World Heritage site helped me put difficult themes into perspective and shed much light on why the islands are now endangered. Carlos Valle, a professor of evolutionary biology at USFQ and GAIAS, who was born in the Galápagos, helped me look deeper into some challenging issues. Günther Reck is director of the Institute of Applied Ecology (ECOLAP) at USFQ and a former director of the Charles Darwin Foundation in the Galápagos. His work early on for the foundation, and with traditional fishermen, opened my eyes to the vicissitudes of change.

The Charles Darwin Foundation was indispensable. Graham Watkins, who served as executive director of the foundation while I lived in the Galápagos, provided me with mountains of data, as did director of operations Felipe Cruz, and communications expert Rosyln Cameron, who was always available to help. My thanks also go to Patricia Robayo, Aldo Jaramillo, Alex Hearn, Gustavo Jimenez, Ivonne Guzman, and Timothy Silcott.

The Galápagos National Park supported my requests for special permission to travel to locations closed to the public. This was critical because I lived there during a chaotic interval at the park headquarters. During most of my time in Puerto Ayora, the park operated with interim leaders because director Raquel Molina was fired. Gratitude goes to René Valle for granting me permission as a National Geographic author to visit Volcán Alcedo on Isabela Island and Punta Cevallo on Española Island; and to Fabián Oviedo; Washington "Wacho" Tapia; Victor Carrión; Lorena Sanchez; and Vanesa García.

Tui de Roy, who grew up in the Galápagos, deserves a bow for putting up with me while hiking up Volcán Alcedo, and for allowing me an intimate view into her life and work as a wildlife photographer. So does Kate Huyvaert, the albatross expert from Colorado State University, and David Anderson from Wake Forest University. Thanks also to Patricia Zárate for allowing me to work in the sea turtle camp at Quinta Playa, and to Mathias Espinosa of Scuba Iguana for help with interviews in Villamil. Alex Cornelissen, director of Sea Shepherd Galápagos, was more than gracious with his time and taught me about the politics of fishing.

Jack Nelson, former owner of the Hotel Galápagos, co-owner of Scuba Iguana, and the U.S. consular agent for the Galápagos, spent hours and hours in interviews, patiently explaining the reality and challenges in the Galápagos during his four decades in Puerto Ayora. His insights and regular columns for the newspaper, *El Colono*, revealed intelligence, perseverance, and a delightful willingness to challenge the system. His sister, Christy Gallardo Nelson, was always an inspiration who shared family and historic photographs, as well as her writing and art.

Abrazos to my naturalist guide friends Fabio Peñafiel and Patricia Stucki, the latter of whom called or text messaged me almost every day just to check up on me when she wasn't out guiding, and to Margarita Brustle, a bright spirit from Austria who I met during her annual visit the Galápagos. Closer to home, many thanks to my friends and colleagues in the United States for supporting me throughout this enormous project: Jack Hart for his early encouragement, Roberta Vanderslice and her husband Dave for unconditional support, my filmmaker colleague Jon Palfreman for keeping me positive, Rick McMonagle for help with publicity, Andrew MacKenzie for designing my author's website, Karen Weitzel for her professional transcription of my interviews, my wonderful family, and the neighbors who looked after my home in Eugene, Oregon, while I was gone: Bill and Donna Eimstad; and Chris, Lauren, and Samantha Murphy. Samantha was three when she helped pull weeds from my overgrown yard.

Finally, I wish to thank my agent Carol Mann in New York, and Barbara Noe, senior editor at National Geographic Travel Books. Barbara's unwavering support for this project—and her belief in me as a writer—gave me enduring strength and a new mantra: *It's all about working towards the light.* I could not have achieved this without her encouragement and insight. Thanks, too, to Susan Hannan for her fine editing.

Following are some tour companies that knowledgeable Galápagos residents recommend: Aggressor I & II Fleet, Andando Cruises, Columbus Travel, Deep Blue Galápagos, Ecoventura, Enchanted Expeditions, Integrity Galápagos Yacht, Kensington Tours, Latin Tour, Lindblad Expeditions, Metropolitan Touring, Ninfa Tour, Scuba Iguana, Tip Top Cruises, Xpedition Cruises, Yacht Daphne, Yacht Darwin, and Yacht Eden.

Bibliography

BOOKS

Beebe, William. *Galapagos: World's End.* New York: G.P. Putnam's Sons, 1924.

Bowlby, John. *Charles Darwin: A New Life.* New York: W.W. Norton & Company, 1990.

Boyce, Barry. *A Traveler's Guide to the Galapagos Islands.* Galapagos Travel, 1990.

Castro, Isabel, and Antonia Philips. *A Guide to the Birds of the Galapagos Islands.* Princeton University Press, 1996.

Chávez, Juan. *Walking Through the Wetlands of Isabela.* Puerto Ayora: Galapagos National Park and USAID, 2004.

Constant, Pierre. *The Galapagos Islands.* Lincolnwood: Odyssey Passport Books, 1997.

———. *Marine Life of the Galapagos: A Guide to the Fishes, Whales, Dolphins, and Other Marine Animals.* New York: W.W. Norton & Co., 2002.

Darwin, Charles. *On the Origin of Species.* New York: Random House, 1979.

———. *Voyage of the Beagle.* New York: Penguin Books, 1989.

———. *Charles Darwin's Beagle Diary.* R.D. Keynes, editor. Cambridge: Cambridge University Press, 2001.

———. *The Descent of Man, and Selection in Relation to Sex.* London: John Murray.

———. Journal of Researches into the Geology and Natural History of the Various Countries Visited by H.M.S. Beagle, Under the Command of Captain FitzRoy, R.N. from 1832 to 1836. London: Henry Colburn.

De Roy, Tui. *Galapagos: Islands Born of Fire.* Toronto: Warwick Publishing, 2004.

Fitter, Julian, Daniel Fitter, and David Hosking. *Wildlife of the Galapagos.* Princeton: Princeton University Press, 2000.

Foreman, Dave. *Earthforce! An Earth Warrior's Guide to Strategy.* Los Angeles: Chaco Press, 1993.

Gordillo, Jacinto. *Guide to Villamil, Isabela, Galapagos.* Puerto Villamil, Ecuador: Municipal Council of Puerto Villamil, 1995.

———. *El Tero Real en Las Islas Galapagos.* Quito: NINA Comunicaciones, 1999.

———. *Relatos de 44 Años en Galapagos,* Quito: Ediciones Abya-Yala, 2000.

Gould, Stephen Jay. *Ever Since Darwin.* New York: W.W. Norton, 1977.

Harris, Michael. *A Field Guide to the Birds of Galapagos.* New York: Collins, 1982.

Heller, Peter. *The Whale Warriors.* New York: Free Press, 2008.

Huxley, Julian. *Evolution: The Modern Synthesis.* New York: Harper & Bros., 1943.

Idrovo, Hugo. *Galapagos: Footsteps in Paradise.* Quito: Ediciones Libri Mundi, 2005.

Jackson, M.H. *Galapagos: A Natural History.* Calgary: University of Calgary Press, 1989.

Larson, Edward J. *Evolution's Workshop: God and Science on the Galapagos Islands.* New York: Basic Books, 2001.

Lyell, Charles. *Principles of Geology.* New York: Penguin Classics, 1998.

McBirney, A.R., and H. Williams. *Geology and Petrology of the Galapagos Islands.* The Geological Society of America, 1969.

McCullen, Conley. *Flowering Plants of the Galapagos.* Cornell University Press, 1999.

Melville, Herman. *The Encantadas and Other Stories.* Mineola: Dover Publications, 2005.

Merlen, Godfrey. *Restoring the Tortoise Dynasty: The Decline and Recovery of the Galapagos Giant Tortoise.* Charles Darwin Foundation and Galapagos National Park, 1999.

Numbers, Ronald L. *The Creationists.* Berkeley: California Press, 1992.

Philbrick, Nathaniel. *In the Heart of the Sea: The Tragedy of the Whaleship Essex,* New York: Viking, 2000.

Preston, Diana, Michael Preston. *A Pirate of Exquisite Mind: Explorer, Naturalist, and Buccaneer: The Life of William Dampier.* New York: Walker & Company, 2004.

Quammen, David, *The Song of the Dodo: Island Biogeography in an Age of Extinctions*. New York: Touchstone, 1997.

_____. *The Flight of the Iguana: A Sidelong View of Science and Nature*. New York: Touchstone, 1998.

_____. *The Reluctant Mr. Darwin*. New York: W.W. Norton, 2006.

Safina, Carl. *Song for the Blue Ocean*. New York: Holt Paperbacks, 1999.

_____. *Voyage of the Turtle*. New York: Holt Paperbacks, 2007.

Swash, A., and R. Stills. *Birds, Mammals and Reptiles of the Galápagos Islands*. Yale University Press, 2006.

Vonnegut, Kurt. *Galápagos* (fiction). Delta Fiction, 1999.

Wallace, Alfred Russel. *Island Life*. New York: Cosimo Classics, 2007.

Weiner, Jonathan. *The Beak of the Finch*. New York: Vintage, 1995.

Whitney, Charles A. *Whitney's Star Finder*. New York: Alfred A. Knopf, 1989.

Wittmer, Magret. *Floreana: A Woman's Pilgrimage to the Galápagos*. New York: Moyer Bell Ltd., 1990.

MAGAZINES AND JOURNALS

Benchley, Peter. "Galapagos: Paradise in Peril." *National Geographic*, April 1999.

Bensted-Smith, Robert. "The War Against Aliens in the Galapagos." *World Conservation*, April 1997.

Bueche, Shelley. "Aggie and Buck: Environmental Sniffers in the Galapagos Islands." *Just Labs*, January-February 2008.

Coblentz, Bruce E. "Strangers in Paradise: Invasive Mammals on Islands." *World Conservation*, April 1997.

De Roy, Tui. "In the Nick of Time: Alcedo's Tortoise Land." *Galapagos News*, a publication of Friends of the Galapagos, Spring 2006.

_____. "Where Giants Roam: Observing Giant Tortoises on Galapagos Island." *Natural History* Magazine, 1997.

Dillard, Annie. "Innocence in the Galapagos." *Harper's*, May 1975.

Emory, Jerry. "Managing Another Galapagos Species—Man." *National Geographic*, January 1988.

Green, Derek. "The East Pacific Green Sea Turtle in Galapagos." *Noticias de Galapagos*, 1978.

_____. "The Green Sea Turtle Project in Galapagos: Past, Present and Future." *Noticias de Galapagos*, 1981.

Huyvaert, Kathryn P., David J. Anderson, Patricia G. Parker, "Mate Opportunity Hypothesis and Extrapair Paternity in Waved Albatrosses (Phoebastria Irrorata)." *The Auk* (American Ornithologists' Union), 123 (2):524-536, 2006.

Koenig, Kevin. "Ecuador's Oil Change: An Exporter's Historic Proposal." *Multinational Monitor*, Sept./Oct. 2007.

Matthiessen, Peter. "In the Dragon Islands." *Audubon*, September 1973.

Meyer, Richard. "First-Hand Report on Galapagos Island Hunt." *The Hunting Report*, May 1999.

Miles, Jonathan. "Great Shark Slaughter." *Men's Journal*, Oct. 2006.

Neville, Tim. "Galapagos Unbound." *The New York Times*, Jan. 8, 2006.

Nelson, Jack. "Four Guys Lobster." Unpublished manuscript, Galápagos, 2008.

Orr, Allen. "Devolution: Why Intelligent Design Isn't." *The New Yorker*, May 30, 2005.

Pearce, Fred. "Galapagos Tortoises Under Siege." *New Scientist*, September 1995.

Pinson, Jim. "Electronic Mail Comes to the Galapagos." *Noticias de Galapagos*, July 1995.

Plage, Dieter, and Mary Plage. "A Century After Darwin's Death: Galapagos Wildlife Under Pressure." *National Geographic*, January 1988.

Quammen, David. "Was Darwin Wrong?" *National Geographic Adventure*, November 2004.

_____. "Return to Zootopia," *National Geographic Adventure*, November 2004.

Randall, Lisa. "Galapagos Reconsidered." *Discover* Magazine, February 2007.

Safina, Carl. "On the Wings of the Albatross," *National Geographic*, December 2007.

Seminoff, Jeffrey A., Patricia Zarate, Michael Coyne, David G. Foley, Denise Parker, Boyd N. Lyon, Peter H. Dutton. "Post-nesting migrations of Galapagos green turtles *Chelonia mydas* in relation to oceanographic conditions: integrating satellite telemetry with remotely sensed ocean data." *ESR Inter-Research*, December 7, 2007.

Shermer, Michael. "The Woodstock of Evolution." *Scientific American*, June 2005.

Talbot, Margaret. "Darwin in the Dock: Intelligent Design Has Its Day in Court." *The New Yorker*, December 5, 2005.

Villiers, Alan. "In the Wake of Darwin's Beagle." *National Geographic*, October 1969.

Watson, Paul. "Captain's Log." *Sea Shepherd Log*, No. 64, Fall/Winter 2006.

Whitty, Julia. "The Fate of the Ocean." *Mother Jones*, March/April 2006.

Wong, Kathleen M. "In Darwin's Wake: The Academy Sets Sail to the Galapagos," *California Wild* (the Magazine of the California Academy of Sciences), Spring 2003.

NEWSPAPERS AND PRESS RELEASES

"Away from Safety of Galapagos, Green Turtle DC235 Dies From Long Line Fishing." Charles Darwin Foundation (CDF), March 5, 2007.

"Darwin vs. Design: Evolutionists' New Battle." *New York Times*, April 8, 2001.

"Demandan expropiación de dos fincas el El Camote." *El Colono*, Primera Edición de Mayo de 2008.

"Ecuador Chases Citizens Off Galapagos to Save Islands." *The Los Angeles Times,* October 8, 2008.

"Ecuador's President Declares Galapagos in Crisis." *El Comercio*, April 11, 2007.

"Ecuador Pressured to Save Galapagos Wildlife," *The Independent* (London), April 1, 2005.

"El Niño's Victims Teeter on Edge of Disaster in Ecuador." *Associated Press*, March 12, 1998.

"Fondo busca eliminar especies invasoras." *El Universo*, June 25, 2007.

"Paradise Lost in the Galapagos Islands." *The Independent* (London), April 18, 2005.

"Fossils Add More Proof of Global Climate Change." *The New York Times*, August 5, 2008.

"Fragile Islands Under Pressure from Nature-Loving Tourists." *The Christian Science Monitor*, August 19, 1991.

"Galapagos Is Added to Endangered List." *The New York Times*, July 8, 2007.

"Galapagos Face 'Two-edged Sword.'" *The Register-Guard*, April 9, 2006.

"Galapagos Islands Face New Peril As More Oil Spills from Tanker." *New York Times*, January 25, 2001.

"Galapagos Sea Lions Butchered." *The Scotsman*, July 19, 2001.

"Galapagos Turmoil: Tortoises Dragged into Fishing War." *The Guardian* (London), December 30, 2000.

"Heat's on Booming Galapagos." *The Los Angeles Times*, October 8, 2008.

"How Graft Is Wearing Away the Galapagos." *Seattle Times*, Jan. 6, 2006.

"Invasive Species Threaten Galapagos Diversity." *Washington Post*, February 27, 2006.

"Las Bajas Notas de los maestros confirman la crisis educativo." *El Comércio*, February 29, 2008.

"Ministra cesó a Directora del Parque Nacional Galapagos." *El Colono*, Secunda Edición de Marzo de 2008.

"Pirate Patrol: Illegal Fishing Threatens the Galapagos." *The Guardian* (London), September 19, 2001.

"Rampaging Galapagos Fishermen Put Islands and Creatures at Risk." *San Francisco Chronicle*, December 10, 2000.

"Sea Shepherd off to Galapagos Islands." *Seattle-Post Intelligencer*, November 26, 2000.

"Sea Shepherd opera en Galapagos con la Policía." *El Comércio*, July 20, 2007.

"The Galapagos Islands: Paradise Lost?" *The Independent*, April 18, 2005.

"Tortoises Held Hostage as Lobster War Turns Nasty." *The Independent* (London), November 18, 2000.

"UNESCO Considers Putting Galapagos on Endangered List." *The New York Times*, April 29, 2007.

"Unnatural Selection: Rampaging Galapagos Fishermen Put Islands and Creatures at Risk." *San Francisco Chronicle*, December 10, 2000.

INTERNET PUBLICATIONS

"1,000 Tortoises Repatriated." www.gct.org, March 16, 2006.

"8 More Eggs Discovered in Lonesome George's Enclosure." www.galapagos.org, August 5, 2008.

"A Legal Threat From Ecuador Over Its Debts." www.nytimes.com, September 27, 2008.

"A Swimming Anthropologist Fights for the Galapagos." *Chronicle of Higher Education,* www.chronicle.com, June 16, 2006.

"Adaptive Radiation of Feral Dogs on the Island of Isabela in the Galapagos." www.stanford.edu, October 12, 2003.

"Addressing Invasive Species in the Galapagos World Heritage Site." www.unfoundation.org, January 7, 2000.

"An Analysis of Nature Tourism in the Galapagos Islands." www.darwinfoundation.org, 2001.

"Artisanal Fishing as a Cultural Experience, a Novel Alternative." www.gct.org, August 17, 2005.

"Artisanal Open Water Fishing an Alternative for the Galapagos Fishing Sector." www.darwinfoundation.org, June 2, 2004.

"Assessing Growth in the Galapagos." www.nature.org, September 2001.

"Authoridades demandan soluciones a problemas de Galapagos." www.isabelagalapagos.com, November 17, 2008.

"Back from Extinction: Saving the Endangered Plants of the Galapagos." www.gct.org, March 16, 2006.

"Biodiversity Conservation and Human Population Impacts in the Galapagos Islands, Ecuador." www.darwinfoundation.org, 1996.

"Biodiversity in the Galapagos." The Galapagos Coalition, www.law.emory.edu.

"Can Extinct Galapagos Tortoise be Bred From Living Hybrids?" www.blogs.nationalgeographic.com/blogs, September 23, 2008.

"Can the Galapagos Survive?" www.time.com, Oct. 30, 1995.

"CDF Studies Reveal Unwelcome Visitors Traveling With Tourist Boats." www.gct.org, May 31, 2007.

"CDRS suffers vandalism and threats of violence at the hands of local fishermen." www.darwinfoundation.org, November 17, 2000.

"Charles Darwin Foundation Names New Executive Director." www.darwinfoundation.org, September 26, 2008.

"Charles Darwin Foundation Warns About Upcoming Experiment with Iron Dust Near the Galapagos Marine Reserve." www.darwinfoundation.org, June 14, 2007.

"Charles Darwin Foundation's Vice President Is the New Governor of Galapagos." www.darwinfoundation.org, August 9, 2007.

"Charles Darwin Research Station, Galapagos, Ecuador." www.darwinfoundation.org, July 1963.

"Civil Unrest Forces Out Ecuador's President." www.en.wikinews.org, April 21, 2005.

"Committee of Concerned Galapagos Citizens." www.naturalist.net, November 21, 2000.

"Community Watchdog Group Established to Observe the Committee for the Qualification and Control of Residency." www.galapagos.org, June 30, 2008.

"The Complete Works of Charles Darwin." www.darwin-literature.com, 2003.

"Conservation on the Brink." www.igtoa.org, January 2005.

"Country Profile: Ecuador." BBC News, www.news.bbc.co.uk, April 16, 2008.

"Curriculum Vitae de Eduardo Alejandro Véliz Veéliz." www.hy.com.ec, June 6, 1997.

"Darwin's finches at risk." www.news.bbc.co.uk, November 8, 2002.

"Darwin's Revolutionary Theory." www.agiweb.org, 2001.

Desmond, Adrian. "The Cautious Evolutionist." www.nytimes.com, August 27, 2006.

"Despite Efforts, Tours Do Leave Footprints." www.washingtonpost.com, April 2, 2006.

"Día Mundial de la Tierra." Ambiente Independiente, www.ambienteindependiente.org, April 2008.

"Did Galapagos Turtle Lineage Survive Ancient Blast?" www.news.nationalgeographic.com, October 2, 2003.

"Economic Overview and History." www.ecuadorexplorer.com, 2007.

"Ecuador appears likely to rewrite constitution." International Herald Tribune, www.iht.com, April 16, 2007.

"Ecuador Congress Sacks President." www.newsvote.bbc.com.uk, April 20, 2005.

"Ecuador Constitution Grants Rights to Nature." www.dotearth.blogs.nytimes.com, September 29, 2008.

"Ecuador Opposes Outpost in American War on Drugs." www.nytimes.com, May 12, 2008.

"Ecuador Referendum Endorses New Socialist Constitution." www.guardian.co.uk, September 29, 2008.

"Ecuador: Oil Rights or Human Rights?" www.amnestyusa.org.

"Ecuador: Protests Paralyze Country." Weekly News Update on the Americas, www.home.earthlink.net/~nicadlw/wnuhome.

"Ecuador's New Constitution Grants Rights to Nature." www.greenchange.org, October 30, 2008.

"Ecuador's Poor Bank on Referendum." www.newsvote.bbc.co.uk, September 27, 2008.

"Ecuador's President Declares Emergency Over Galapagos Oil-Spill Threat." www.cnn.com, January 22, 2001.

"Ecuador's President Gets No Slack." www.commondreams.org, May 1, 2005.

"Ecuador's President Vows to Focus on Poor." www.boston.com, January 16, 2007.

"Ecuadoreans back new constitution." www.bbc.co.uk, September 29, 2008.

"Effects of the 1997-98 El Niño event on the vegetation of the Galapagos." www.darwinfoundation.org, December 1999.

"Effects of Tourism: Observations of a Resident Naturalist." Noticias de Galapagos, www.darwinfoundation.org, 1980.

"Environmental Stewardship." Scuba Iguana, www.scubaiguana.com, 2002.

"Environment-Ecuador: A Day of Mourning for Galapagos Islands." www.oneworld.org, March 2, 1998.

"The Era of Catastrophe? Geologists Name New Era after Human Influence on the Planet." www.alternet.org, August 11, 2008.

"The Evolution of Corruption." www.csmonitor.com, April 3, 2008.

"The Evolution of Galapagos Wind." www.galapagos.org, March 2006.

"Exotic Sky Adventures: Next Event." www.exoticskyadventures.com, 2006.

"Exploring Ocean Iron Fertilization: The Scientific, economic, legal and political basis." Woods Hole Oceanographic Institution, www.whoi.edu, 2007.

"Fernandina." www.geo.mtu.edu/volcanoes, 1995.

"Fishermen Linked to Decline in Galapagos Waved Albatross Population." www.gct.org, May 31, 2007.

"Fishermen Strike in Galapagos." www.wildaid.org, March 1, 2004.

"Fishing Imperils Galapagos: Ecuador Islands face delisting by United Nations." www.vancouversun.com, April 7, 2005.

"Fundacion Natura, Ecuador: Institutional Mission." www.latinsynergy.org.

"The Galapagos: WWF Delivers Lasting Results." www.worldwildlife.org, 2006.

"Galapagos Flora." www.gct.org, September 13, 2008.

"Galapagos Geology on the Web." www.geo.cornell.edu/geology.

"Galapagos Islands control tourism and immigration with secure plastic identification cards and Zebra P420i printers." www.galapagos.org, September 28, 2006.

"Galapagos Islands Face New Peril as More Oil Spills from Tanker." www.nytimes.com, January 25, 2001.

"Galapagos Islands Journal; Homo Sapiens at War on Darwin's Peaceful Isles." www.nytimes.com, November 28, 1995.

"Galapagos Marine Biodiversity Fund." www.ecoventura.com, 2008.

"Galapagos Marine Patrols Gain Speed on Seafaring Outlaws." www.worldwildlife.org, October 30, 2006.

"Galapagos Must Evolve." www.darwinfoundation.org, 2006.

"Galapagos Petrel." www.darwinfoundation.org, 2006.

"Galapagos to Be Only Feral Goat-Free Archipelago in the World by 2010." www.galapagos.org, September 22, 2008.

"Galapagos Tortoises Under Siege." www.newscientist.com, September 16, 1995.

"Galapagos Unbound." www.nytimes.com, January 8, 2006.

"Galapagos: Where Ridge Meets Hotspot." www.oceanexplorer.noaa.gov, December 3, 2005–January 10, 2006.

"Giant Tortoises Still on the Menu—4 More Slaughtered in Isabela." www.darwinfoundation.org, September 26, 2006.

"Gold mine bug 'lives without light, oxygen.'" www.abc.net.au/news, October 11, 2008.

"Hammerhead Shark." www.en.wikipedia.org, March 2008.

"Hotel Galapagos History." www.hotelgalapagos.com.

"Hyperbaric Medicine." www.en.wikipedia.org.

"Illegal Fishing Boat Captured in Galapagos Marine Reserve." www.galapagos.org, September 26, 2008.

"Institute for Creation Research." www.en.wikipedia.com, July 27, 2006.

"Jorge Torres Named the New Governor of Galapagos." www.galapagos.org, October 1, 2008.

Khan, Stephen. "Shark-fin soup eaters putting Galapagos ecosystem at risk." CDNN (Cyber Diver News Network), www.cdnn.info/news/eco, 2007.

"La Polémica gira en torno a la destitución de Raquel Molina." www.elcomercio.com, March 11, 2008.

"The Land Iguana: Back from the Brink." www.gct.org. March 29, 2007.

"Let 'intelligent design' and science rumble." www.latimes.com, October 2, 2005.

"Lonesome George's Path in Search of Paternity." www.darwinfoundation.org, July 30, 2008.

"Longline Campaign." www.seashepherd.org, 2008.

"Longline Fishing News." www.igtoa.org, April 11, 2006.

"Looking Back on Twenty Years of the Charles Darwin Foundation." www.darwinfoundation.org, 1979.

"Making Waves on the Galapagos." *Chronicle of Higher Education*, www.chronicle.com, June 16, 2006.

"Marine Conservation and Human Conflicts in the Galapagos Islands." www.depts.washington.edu, 2000.

"Methodology of Project Isabela in the Galapagos Islands." www.gct.org, April 17, 2006.

"Mi Poder en la Constitución." www.ecuaworld.com, 2008.

"Milestones in the History of Galapagos and the Charles Darwin Foundation." www.darwinfoundation.org, 2005.

"Military Officers Fired Over Fracas with Galapagos Park Staff." www.news.notiemail.com, July 31, 2007.

"Molecular Science Comes to the Galapagos." www.scidev.net, August 8, 2003.

"Molina dejó ayer la Dirección del PNG." www.elcomercio.com, March 7, 2008.

O'Hearn, Sean. "Sea Shepherd Galapagos Sting Results in Seizure of Over 19,000 Shark Fins." www.seashepherd.org, June 2007.

"Oil Spill Leads to Large Scale Dieoffs in Galapagos." www.darwinfoundation.org, June 7, 2002.

"On the Origin of Revolution." www.newscientist.com, September 30, 1995.

"Operation Aquatic Iron Dust Storm." www.seashepherd.org, July 18, 2007.

"The Origin of Species—Creationism and Evolution." www.hawksbill.com.

"Our Galapagos Commitment." Ecoventura, www.ecoventura.com.

"Pink Land Iguanas Classified as Separate Species in Galapagos." www.igotoa.org, January 6, 2009.

"Pinta Giant Tortoise." www.darwinfoundation.org, 2006.

"Pirates of the Galapagos: British submarine seized." www.news.independent.co.uk, July 9, 2006.

"Planet's Loneliest Bug Revealed." www.news.bbc.co.uk, October 10, 2008.

"Planktos Is a No Show in the Galapagos." www.seashepherd.org, August 10, 2007.

"Political Parties are most corrupt institutions worldwide, according to TI Global Corruption Barometer 2004." Transparency International, www.transparency.org/policy, December 9, 2004.

"Population Status of the Critically Endangered Waved Albatross Phoebastria irrorata, 1999 to 2007." www.int-res.com, May 28, 2008.

"Position of Fundación Natura Before the Alarming situation of Artisan Fishing in the Galapagos Islands." www.babel.altavista.com, November 17, 2000.

"President Wins Support for Charter in Ecuador." www.nytimes.com, September 29, 2008.

"Project Isabela Achieves the Impossible." www.darwinfoundation.org, July 5, 2006.

"Project Isabela," www.galapagos.org, 2008.

"Puerto Ayora Journal; Galapagos Burden: Goats, Pigs and Now People." www.nytimes.com, September 30, 1993.

"Pumping iron into ocean gets chilly reception." www.miamiherald.com, October 7, 2007.

"Research Station Presents Excellence Award to Naturalist Guides of Galapagos." www.darwinfoundation.org, July 2000.

"Saving Endangered Species: Captive Breeding, Rearing, and Repatriation." www.galapagos.com, 2008.

"Saving the Galapagos from Disneyism." www.projo.com, March 23, 2008.

"Saving the Galapagos." www.nytimes.com, October 12, 1995.

"Scientists assess diesel spill impacts in Academy Bay." www.fcdarwin.org, September 16, 2005.

"Scientists discover volcanic activity in the Galapagos with the aid of satellite radar." www.stanford.edu, October 20, 2000.

"Sea Cucumber Loss in the Galapagos." www.american.edu.

"Sea Cucumbers." *Charles Darwin Foundation*, 2006.

"Sea Lion Monitoring and Conservation." www.darwinfoundation.org.

"Sea Shepherd Galapagos Unveils Investigation of Political Corruption." www.seashepherd.org, June 12, 2007.

"Sea Shepherd Helps Establish Permanent Base at Darwin and Wolf." www.seashepherd.org, October 17, 2008.

"Sea Shepherd in the Galapagos." www.seashepherd.org, 2006.

"Sea Shepherd Investigation Leads to Indictment of Mangrove Destroying Mayor." www.seashepherd.org, July 10, 2007.

"Sea Shepherd Partners with World Wildlife Fund to Protect the Galapagos." www.seashepherd.org, October 31, 2006.

"Sea Shepherd Welcomes Raquel Molina as the new Director of the Galapagos National Park." www.seashepherd.org, May 10, 2006.

"Seabirds Give New Meaning to Sibling Rivalry." National Science Foundation, www.nsf.gov, June 1997.

"Shark Campaign." www.gct.org, February 22, 2008.

"Shark Fin Soup: An Eco-Catastrophe?" by Hank Pellissier, www.sfgate.com, January 20, 2003.

"Sharks from the Galapagos Marine Reserve Migrate to Other Zones in the Pacific." www.galapagos.org, November 20, 2007.

"Sharks in Galapagos." www.darwinfoundation.org, 2007.

"Showdown in the Galapagos." www.seawifs.gsfc.nasa.gov.

"Siblicide in Nature: Study of Galapagos Seabird Finds Death Can Ensure Species Survival." www.scienceblog.com/community, Wake Forrest University, 2004.

"Sierra Negra Volcano Erupts." www.galapagos.org, October 22, 2005.

"Skeptic: Extraordinary Claims, Revolutionary Ideas, and the Promotion of Science." www.skeptic.com, July 28, 2006.

"Slaughter of Sea Lions Condemned by Conservationists." www.gct.org, March 16, 2006.

"Smelly Fishy Business in the Galapagos." www.seashepherd.org, May 8, 2008.

"The Special Law for Galapagos." www.darwinfoundation.org, from *Noticias de Galapagos*, April 1998.

"Sport Fishing Update." International Galapagos Tour Operators Association, www.igtoa.org, June 2006.

"Study of the U.S. Air Forces Galapagos Islands Base." www.galapagos.to.

"The Survival Strategies of Some Threatened Galapagos Plants." www.darwinfoundation.org, from *Noticias de Galapagos*, 1979.

"Tanker Captain, Crewmen Arrested in Galapagos Fuel Spill." www.cnn.com, January 25, 2001.

"That First Iguana Transfer." www.galapagos.to, 2002.

"Thoreau's Legacy: American Stories about Global Warming." Union of Concerned Scientists, www.ucsusa.org, 2008.

"Three New Eggs Found in Lonesome George's Corral." www.galapagos.org, September 8, 2008.

"Top Ten Burning Issues in Global Sea Turtle Conservation." IUCN/SSC Marine Turtle Specialist Group, www.seaturtlestatus.org, August 2005.

"UNESCO Mission confirms threat to Galapagos Islands." www.whc.unesco.org, April 16, 2007.

"The Unorganized Growth of Tourism Threatens the Future of Galapagos and Leaves Limited Benefits for the Local Population." www.galapagos.org (from a press release by www.cedenma.org), November 15, 2005.

"Urgent: E Alert from the Cofan Territories." www.cofan.org. October 15, 2008.

"Viagra leaves limp demand for illegal impotence remedies, researchers say." www.biopsychiatry.com, December 27, 2002.

"Volcanoes of the Galapagos." www.galapagosonline.com, 2008.

"Voters in Ecuador Approve Constitution." www.washingtonpost.com, September 29, 2008.

"War in the Pacific: It's Hell, Especially if You're a Goat." www.nytimes.com, May 1, 2007.

Watson, Paul. "Hominid Exotics Removed from the Galapagos." www.seashepherd.org, October 9, 2008.

"Waved Albatross—Birdlife Species Factsheet." BirdLife International, www.birdlife.org, 2008.

"We're all arrested in a mini-submarine as Galapagos trip takes a dive." www.nzherald.co.nz, July 10, 2006.

"What is a Hotspot?" www.oceanexplorer.noaa.gov, December 2005.

"What is the Mid-Ocean Ridge?" www.oceanexplorer.noaa.gov, December 2005.

"Who Killed the Iguanas?" www.galapagos.to, 2002.

"Why Do Sharks Have Two Penises?" www.elasmo-research.org.

"Wildaid Responds to Fishermen's Protests." www.wildaid.org, June 1, 2005.

"Will Ocean Iron Fertilization Work?" *Oceanus Magazine*, www.whoi.edu, Feb. 17, 2009.

"Wind Power Blows through Galapagos." www.galapagos.org, February 18, 2008.

"Wind Power Project for Baltra and Santa Cruz." www.galapagos.org, July 25, 2008.

"The Woodstock of Evolution." www.sciam.com, June 27, 2005.

"World Heritage in Danger." www.whc.unesco.org, May 2007.

REPORTS AND STUDIES

Charles Darwin Foundation Annual Report 2006. Charles Darwin Foundation, October 2007.

Charles Darwin Foundation Strategic Plan 2006-2016. Charles Darwin Foundation, 2006.

Compromiso: Hacia la comunidad y el medio ambiente. Fundacion Galapagos (Metropolitan Touring), May 2008.

Epler, Bruce. *Tourism, the Economy, Population Growth, and Conservation in Galapagos.*

Hearn, Alex. *Fact Sheet: Use of Longline in the Galapagos Marine Reserve,* Charles Darwin Research Station, January 2005.

_____. *The Rocky Path to Sustainable Fisheries Management and Conservation in the Galapagos Marine Reserve.* Charles Darwin Foundation.

Informe Galapagos 2006-2007. Galapagos National Park, Charles Darwin Foundation, Ingala. 2007.

Galapagos Report 2006-2007. Charles Darwin Foundation, Galapagos National Park Service, the National Galapagos Institute, 2007.

The Galapagos Marine Resources Reserve Decree, the Official Register of Ecuador, No. 434, May 13, 1986.

The Thematic Atlas of Project Isabela. Charles Darwin Foundation, et al., 2007.

Una Visión de Futuro Para Galapagos: El Plan de Manejo Para Todos. Parque Nacional Galapagos, 2007.

Watkins, G., and F. Cruz, *Galapagos at Risk: A Socioeconomic Analysis.* Charles Darwin Foundation, May 2007.

AUDIO

ABC News. "Vandals Attack Galapagos Islands." September 12, 2000.

BBC World Service. "Evolution Questioned by Today's Christian Sects on Galapagos Islands." October 5, 2005.

National Public Radio (NPR). "Talk of the Nation: Fishermen, Conservation and the Galapagos." April 16, 2004.

_____. "Fishermen, Conservationists Clash in Galapagos Islands." November 29, 2000.

_____. "Kansas Moves Closer to Intelligent Design Curriculum." August 11, 2005.

_____. "Scientists Hesitant to Debate Intelligent Design." July 8, 2005.

_____. "Remembering the Scopes Monkey Trial." July 5, 2005.

_____. "Darwin Descendant Reflects on Attacks on Evolution." May 12, 2005.

_____. "The Evolution of Evolution Theory." May 21, 2004.

_____. *National Geographic Society Radio Expeditions,* 2-part series, "El Niño in the Galápagos," 1998.

VIDEO

Charles Darwin, BBC Television Film Series.

Evolution, PBS series.

"*Galápagos* IMAX."

"Galapagos," BBC video production.

Judgment Day: Intelligent Design on Trial, Nova series (PBS).

The Blue Planet, BBC series.

Two Years in Galápagos, Australian Broadcasting Corporation.

Voyage to the Galápagos, PBS Scientific American Frontiers.

BROCHURES AND PAMPHLETS

"Bongo Galapagos Bar: Free Body Shots," 2007.

"Ecoventura: Your Guide to the Galapagos," 2005.

"GAIAS: The Galapagos Academic Institute for the Arts and Sciences," 2006.

"GalapagosForever." WildAid, n.d.

"Vientos de Cambio: Proyecto Eólico San Cristobal, Galapagos," March 2008.

MAPS

"Cartografía Galapagos 2006: Conservación en otra Dimensión." The Nature Conservancy, 2006. (Includes CD.)

"Ecuador: Mapa Turístico," 2006.

"Galapagos Islands (Ecuador) Scale 1:500,000," 2004.

"GPS-Guided Galapagos," June 2007–November 2007.

"Islas Galapagos," 2006.

"Isla Isabela," n.d.

"Isla Santa Cruz," n.d.

"Puerto Villamil," n.d.

IMPORTANT WEBSITES

Charles Darwin Research Station
www.darwinfoundation.org

Galapagos National Park
www.galapagospark.org

Galapagos Academic Institute
for the Arts & Sciences (GAIAS)
www.usfq.edu.ec/GAIAS

Universidad San Francisco de Quito
www.usfq.edu.ec

World Wildlife Fund
www.worldwildlife.org

Galapagos Conservation Trust
www.gct.org

Galapagos Conservancy
www.galapagos.org

Conservation International
www.consevation.org

WildAid
www.wildaid.org

Sea Shepherd Conservation Society
www.seashepherd.org

Galapagos Coalition
www.law.emory.edu/PI/GALAPAGOS/

Friends of the Galapagos
www.gct.org/fogos

The Writing of Charles Darwin on the Web
pages.britishlibrary.net/charles.darwin

AboutDarwin.com
www.aboutdarwin.com

About the Author

Carol Ann Bassett is the author of *A Gathering of Stones: Journeys to the Edges of a Changing World*, a finalist for the 2003 Oregon Book Award in Creative Nonfiction, and *Organ Pipe: Life on the Edge* (Desert Places series). Her essays have been anthologized in the *American Nature Writing* series and in *The Mountain Reader: A Nature Conservancy Book*. Bassett was a regular contributor to *The New York Times* and *Time-Life*, Inc. Her work has appeared in *The Nation, Mother Jones, The Los Angeles Times, Condé Nast Traveler, Science 86*, on National Public Radio, and in numerous other national publications. Bassett's narratives focus on natural history and traditional cultures. She teaches at the University of Oregon in Eugene and enjoys kayaking in coastal estuaries, canoeing in the Cascades, hiking, backpacking, and organic gardening.